AF347533

Production and Role of Molecular Hydrogen in Plants

Production and Role of Molecular Hydrogen in Plants

Editor

John T. Hancock

MDPI • Basel • Beijing • Wuhan • Barcelona • Belgrade • Manchester • Tokyo • Cluj • Tianjin

Editor
John T. Hancock
Department of Applied Sciences,
University of the West of England,
Bristol BS16 1QY, UK

Editorial Office
MDPI
St. Alban-Anlage 66
4052 Basel, Switzerland

This is a reprint of articles from the Special Issue published online in the open access journal *Plants* (ISSN 2223-7747) (available at: https://www.mdpi.com/journal/plants/special_issues/Molecular_Hydrogen).

For citation purposes, cite each article independently as indicated on the article page online and as indicated below:

LastName, A.A.; LastName, B.B.; LastName, C.C. Article Title. *Journal Name* **Year**, *Volume Number*, Page Range.

ISBN 978-3-0365-5097-8 (Hbk)
ISBN 978-3-0365-5098-5 (PDF)

Contents

About the Editor

John T. Hancock

John T. Hancock is Professor of Cell Signalling, at the University of the West of England, Bristol, UK. He has had a long-standing interest in the role of reactive oxygen species, nitric oxide, and hydrogen sulphide in biological systems. All these compounds impinge on cellular redox, which is an underpinning theme of his research. Recently, he has been investigating the action of molecular hydrogen in biology, publishing on a range of topics, including the use of H2 for the treatment of COVID-19 and in agriculture.

Preface to "Production and Role of Molecular Hydrogen in Plants"

It seems like the time is right for the discussion of molecular hydrogen (hydrogen gas; H_2) in biological systems. H_2 is rapidly gaining prominence in the scientific literature as well as in the popular media. Therefore, looking at the effects of H_2 in plants seems like a sensible thing to do.

In the biomedical arena, early pre-clinical and clinical studies suggest the use of H_2 treatment for a wide range of human diseases, from COVID-19 to various neurodegenerative diseases. H_2 can be administered as a gas, or it can be given as a solution saturated with H_2, usually referred to as hydrogen-rich water (HRW). Both treatments appear to be beneficial to humans, and it has even been suggested as a sports supplement.

It is no great surprise, therefore, that H_2 has effects in plants too. This is exemplified by the suggestion that H_2 will be useful in large-scale agricultural settings. Such use is supported by the findings so far. H_2 has effects on a range of physiological events in plants. It has been shown to have effects on seed germination, plant growth, and development. It has also been found to be involved in plant stress responses and to be protective against pathological abiotic stress challenges. Similarly, it has also been shown to have beneficial effects during the post-harvest storage of crops, including flowers. Therefore, its use in the agricultural setting has great potential. H_2 not only appears to have likely benefit but it also seems to be to be safe, with no toxicity or harm to the environment.

One of the conundrums of the use of H_2 is understanding how it induces the observed effects in plants and plant cells. It is difficult to envisage how it works based on a classical receptor mechanism. There is some evidence that it may act as a direct antioxidant, especially by scavenging hydroxyl radicals, or via enhancing the plant's antioxidant system as a biological signalling molecule. It has also been reported to have some effects through its possible action on heme oxygenase, crosstalk with other signalling molecules (especially reactive oxygen and nitrogen species) and regulating the expression of various genes. However, how molecular hydrogen fits into, and integrates with, other signalling pathways is not well understood. Future work will no doubt help to elucidate the mechanism(s) and significance of the interaction of H_2 with these and other cellular systems.

This Special Issue aimed to bring together a body of papers that focus on the current state-of-play of the molecular biology and possible uses of H_2 with plants. It is hoped that this Special Issue has highlighted elements of future work which may be undertaken in this field. It is also hoped that on reading these papers, researchers will be encouraged to investigate this exciting field further, bringing new ideas and enthusiasm to the subject.

Lastly, I would like to thank all the people involved in pulling this body of work together. This includes the authors of the papers, the researchers who kindly refereed the papers and all the team at MDPI, especially Snow Liu, who has been working tirelessly in the background making all this happen.

John T. Hancock

Editor

Editorial

Editorial for Special Issue: "Production and Role of Molecular Hydrogen in Plants"

John T. Hancock

Department of Applied Sciences, University of the West of England, Bristol BS16 1QY, UK; john.hancock@uwe.ac.uk; Tel.: +44-(0)1173282475

Citation: Hancock, J.T. Editorial for Special Issue: "Production and Role of Molecular Hydrogen in Plants". *Plants* **2022**, *11*, 2047. https://doi.org/10.3390/plants11152047

Received: 15 July 2022
Accepted: 3 August 2022
Published: 5 August 2022

Publisher's Note: MDPI stays neutral with regard to jurisdictional claims in published maps and institutional affiliations.

1. Introduction

Molecular hydrogen (H_2) is an extremely small molecule, which is relatively insoluble in water and relatively inert. Regardless of this, there seems little doubt that H_2 has profound effects in a range of organisms, from plants [1] to humans. In the biomedical arena, H_2 has been suggested as a therapeutic agent [2] for a range of diseases, including neurodegenerative disease [3] and COVID-19 [4]. It has even been suggested to be useful as a sports supplement [5].

H_2 can be administered to plants, or plant tissues, either as a gas or dissolved in a suitable medium and sprayed on. For the latter, H_2 is usually bubbled through water to make a hydrogen-enriched solution, referred to as hydrogen-rich water (HRW). This can then be added to the soil (or feed solution) or sprayed on the foliage. As an example, Wu et al. [6] used this approach to look at effects of H_2 on cadmium stress in cabbage.

Using such approaches there have been numerous reports of the effects of H_2 in plants. H_2 is involved in seed germination, for example, especially under salt stress [7]. Hydrogen-rich water (HRW) promotes root growth, again especially under stress conditions [8], such as when excess metal ions are present. It has been suggested that such stress relief by H_2 may involve phytohormone signalling [9].

H_2 is safe. Sun et al. [10] give three reasons why they consider H_2 safe for humans, and, therefore, by extension it is safe for treatments of plants, which are used for food crops. Firstly, H_2 has been used as a compressed gas for deep-sea diving for decades, with no ill effects reported. Secondly, H_2 is an endogenous gas, being produced in the gut, for example [11]. Thirdly, experimental evidence has been reported that H_2 is safe. On the other hand, H_2 is highly inflammable, so some caution needs to be exercised if used in confined places, such as in a glasshouse.

How H_2 is acting on plants is not well understood. If sprayed onto the foliage it is not known how much of the H_2 enters the plant tissues but clearly it has to if there are effects seen. Even if used as HRW, H_2 is very likely to enter the gas phase relatively quickly, so repeated treatments may be needed. The direct effects of H_2 are also hard to understand, i.e., what the molecular targets of H_2 are. H_2 is probably too small to interact with a protein receptor in the classical manner. However, there are reports of H_2 acting as an antioxidant [12], with the most likely target being the hydroxyl radical, or the peroxynitrite molecule ($ONOO^-$). The latter is formed through a reaction of nitric oxide (NO) and superoxide anions. It is thought that it is less likely that H_2 is reacting with other reactive oxygen species (ROS), such as the superoxide anion or hydrogen peroxide (H_2O_2), or other reactive nitrogen species, such as NO. It is also unlikely that H_2 is involved in the direct modification of amino acids in polypeptides, which would be the typical mechanism of H_2O_2 (through oxidation [13]) or of NO (through S-nitrosylation [14]). Others have suggested that H_2 acts on enzymes, such as heme oxygenase [15], but the direct interaction is not known.

Here, this Special Issue (SI) was an invitation to those in the field to give their up-to-date research and appraisal of the effects and uses of H_2 in plant science.

2. Effects of H$_2$ on Plants

Here, in this SI, Nguyen and Lim ask if H$_2$ can be used for the extension of the vase life of flowers [16]. Post-harvest is a hugely important topic. In some countries, the postharvest waste has been estimated to be over 40% [17]. Crops need to be commercially viable, safe to consume and acceptable to the customers. This latter point is very pertinent for flowers where the expectation would be that they look good and last for as long as possible. Nguyen and Lim reviewed the various methods used to deliver H$_2$ to plants, including HRW, hydrogen nanobubble water (HNW) and magnesium hydride (MgH$_2$). Plants used across studies included rose, carnation and lilies. Overall, the application of H$_2$ increased vase life of flowers, and the authors concluded that such work should be continued to release the potential for the use of H$_2$ in the floriculture industry. Interestingly, the authors also discuss the cost–benefit analysis of H$_2$ use and suggest that labour costs are an issue [18]. Certainly, as H$_2$ is developed more for other industries, such as a transport energy source [19], the cost and delivery of H$_2$ is likely to become cheaper.

Li et al. [20] continue the theme of using H$_2$ to extend the vase life of flowers. Here, the authors use HNW, which they say has a higher concentration of dissolved H$_2$ than conventional HRW, and the residual time that the H$_2$ remains dissolved for is longer, both properties, which would be of benefit if adopted for widespread use. Their data show that 5% HNW significantly lengthened the vase life of cut carnations. This concentration was better than other concentrations of HNW tried, and better than either water or HRW. Their measures of improvements included electrolyte leakage, oxidative damage and cell death in the petals. The authors concluded by suggesting that HNW may have future applications for postharvest preservation. Certainly, treatment with molecular hydrogen is likely to be much safer than the use of some of the alternatives mooted, such as hydrogen sulfide [21], which is known to have toxicity [22]. This might not be too much of an issue with flowers but may become a problem if the same treatments are used for postharvest preservation of food crops.

Cheng et al. [23] have taken H$_2$ applications out into the field, with trials of whether such treatments will improve rice. They too use HNW, comparing it to ditch water. Their data show that HNW increased the length, width and thickness of brown/rough rice and white rice. They then looked at gene expression in these plants and could correlate the physiological changes with the molecular alterations seen. In the white rice they saw no difference in total starch content, but the enzyme amylase was decreased. Cadmium accumulation was also decreased, which also correlated with gene expression patterns reported. Overall, such work shows that H$_2$ treatments can be taken to larger scales, not just in a laboratory setting. The authors conclude that H$_2$ application does increase the quality of the rice and should be considered as a future treatment. Although field trials with H$_2$ have been written about before [24], they are relatively rare and more large-scale work, such as this, certainly needs to be undertaken if H$_2$ treatments are to be used more widely in agriculture.

There seems little doubt that H$_2$ treatment of plants is beneficial, as exemplified by the papers in this SI and the further papers that these authors cite. There is a growing body of this evidence, and as more is reported, on different species, the use of H$_2$ will be seen to be advantageous, whether used in the field or postharvest. Despite the molecular mechanisms not being well understood, the next papers go someway to unravel what H$_2$ might be doing in plant tissues.

3. Mechanisms of H$_2$ Action

The mechanisms in the cells, which enable H$_2$ to have its effects, were also the focus of papers in this SI. Zhao et al. [25] show an interaction between H$_2$ and glucose in adventitious roots of cucumber. The effects of HRW were blocked by glucosamine, suggesting that glucose content may be mediating root development. HRW increased the cellular content of a range of sugar-based metabolites, including glucose, starch, sucrose, glucose-6-phosphate, fructose-6-phosphate and glucose-1-phosphate. HRW treatment resulted in the increase in the activity of several relevant enzymes, including hexokinase, pyruvate kinase and sucrose synthase, and, furthermore, gene expression patterns matched these findings. Interestingly, the authors state that all the positive effects of HRW were inhibited by glucosamine, and they concluded that H_2 was regulating adventitious root growth by promoting glucose metabolism.

The literature on H_2 effects tends to support the notion that H_2 increases cellular antioxidant levels. For example, Wu et al. [6] reported increased antioxidants in cabbage under cadmium stress, whilst Chen et al. [26] showed that the antioxidant capacity of *Hypsizygus marmoreus* (mushroom) was increased by HRW use during postharvest. Here, in this SI, Jiang et al. [27] also look at antioxidant capacity and how this might be improved by H_2 treatment, using Chinese chive. In a similar manner to the vase-life work above, this is also being carried out postharvest. Chives were treated with a range of H_2 concentrations, in comparison to air. Shelf-life was improved most by 3% H_2, a conclusion supported by measurements of decay index, loss ratio of weight and protein content. Of pertinence to the discussion here, the content of total phenolics, flavonoids and vitamin C were maintained, whilst the activities of antioxidant enzymes were increased, including superoxide dismutase (SOD), catalase (CAT) and ascorbate peroxidase (APX). Clearly, H_2 treatment was protecting the plant material by an increase in antioxidant capacity, as suggested by others, for example [28].

Zhang et al. [29] looked at pesticide residues in plants. Using tomato and Arabidopsis, they found that the degradation of carbendazim (a benzimidazole pesticide) was increased by H_2. H_2 increased glutathione metabolism, which led to the increased degradation of carbendazim. Glutathione is an immensely important antioxidant in plants [30], so any effects seen may also alter cellular redox states [31]. Carbendazim is a fungicide [32]. Zhang et al. [29] comment that the antifungal action of carbendazim was not affected by the H_2 treatments, but they were convinced that glutathione was important for the H_2 detoxification of the fungicide. Therefore, as the authors point out, this is a previously unknown action of H_2 in plant cells.

Wuzhimaotao (*Ficus hirta Vahl*) was the plant of focus for the study by Zeng and Yu [33]. In China, this edible plant also has medicinal properties. Following treatment with H_2 (as HRW), the transcriptomic pattern of the roots was compared with controls. One hundred and seventy three genes were found to be down regulated, whilst 138 were up regulated. The authors also carried out a metabolomic analysis and found that nearly 200 metabolites had their levels altered by H_2 treatment. With further analysis, it was suggested that the biosynthesis and metabolism of phenylpropanoid were the main pathways being controlled by H_2 treatment. Although the authors suggest that H_2 application should be considered for future growth of this medicinal plant, the data also show the scale of the effects that can be reported when plants are treated with H_2. With so many genes being up- and down-regulated, H_2 must be having a profound effect on transcription factors in plant cells. With similar results emanating from work on mammalian cells [34], the future will no doubt allow the mechanisms behind these changes in gene expression to be unravelled.

Finally, a review of the direct actions of H_2 on plants was published as part of the SI [35]. There are reports that H_2 acts as a direct antioxidant [12], but the chemistry has been disputed [36]. It has been suggested that H_2 acts through its redox state [37], and there is some precedent for this in bacterial systems [38]. Alternatively, H_2 may be acting through its spin states [39], but there is no evidence of this. Clearly, there is a lot to explore here, and some of the focus of H_2 research needs to be pointed in this direction.

Whatever the mechanism, it would strengthen the argument for the use of H_2 in agriculture, and in the biomedical arena, if the mechanism(s) of the direct action of H_2 was resolved, but there is no doubt that such evidence, either ruling mechanisms in or out, will be forthcoming in the future.

4. Conclusions and Future Perspectives

As can be seen by the papers that were published in this Special Issue, H_2 has beneficial effects on plants. H_2 appears to increase the crop yields and can be used for improving postharvest storage of crops, as exemplified by the work on flowers here. Therefore, H_2 use should have a bright future in agricultural settings.

Plants may be exposed to H_2 naturally, either through the action of cellular enzymes [40], or through the metabolism of other organisms in the location, such as in the soil [41]. Alternatively, H_2 may be applied to the plant—either onto the soil or foliage—as a treatment. As can be seen in the papers in this SI, there are a variety of ways to achieve this. H_2 can be applied as a gas, or in an enriched solution, i.e., HRW. However, more recent advances in this area have seen the development of other solutions, which can be used, such as HNW. Alternatively, H_2 can be supplied from donor molecules, such as MgH_2, and no doubt the future will see better donor compounds being developed, which can deliver more H_2 for a longer period of time, rather than giving tissues a bolus effect.

One aspect of the reporting of the effects of H_2 that needs to be consistent is the quoting of the concentration of H_2 used. Often the percentage of HNW or HRW is quoted, but without knowing for sure exactly how concentrated the stock solution is it is hard to compare different studies and, therefore, the effects. H_2 does not last long in solution, so quoting the actual concentration of H_2 in the solutions used would be very beneficial to push this field forward.

Some evidence of the molecular aspects are presented here in this SI too, including the action through glucose metabolism, glutathione metabolism, as an antioxidant and in the control of gene expression. However, the direct targets of H_2 still remain elusive, not just in plant science but in all aspects of the action of H_2 in biological systems. Several ideas have been mooted but there is little evidence of them at present. Future work needs to be focussed on this aspect of H_2 biology, as this would really strengthen the argument for H_2 use. It would also give reassurance on the safety of H_2 treatments, especially if it is proposed to be used as a treatment for consumed crops, either in the field or postharvest.

Finally, more large field trials are needed on a range of crops. H_2 is being studied in some countries around the world, most notably China, but it needs to be looked at more widely, in different locations and with different plants. H_2 appears to be safe, albeit inflammable, but the cost–benefit needs to be well established before H_2 will be taken up widely in agriculture and floriculture. H_2 has benefits, especially if used when plants are stressed, and no doubt large scale trials will unlock the hesitation for the adoption of H_2 applications in the future.

Funding: This research received no external funding.

Conflicts of Interest: The author declares no conflict to interest.

References

1. Zulfiqar, F.; Russell, G.; Hancock, J.T. Molecular hydrogen in agriculture. *Planta* **2021**, *254*, 1–14. [CrossRef] [PubMed]
2. Ge, L.; Yang, M.; Yang, N.N.; Yin, X.X.; Song, W.G. Molecular hydrogen: A preventive and therapeutic medical gas for various diseases. *Oncotarget* **2017**, *8*, 102653. [CrossRef]
3. Ohno, K.; Ito, M.; Ichihara, M.; Ito, M. Molecular hydrogen as an emerging therapeutic medical gas for neurodegenerative and other diseases. *Oxidative Med. Cell. Longev.* **2012**, *2012*, 353152. [CrossRef] [PubMed]
4. Alwazeer, D.; Liu, F.F.C.; Wu, X.Y.; LeBaron, T.W. Combating oxidative stress and inflammation in COVID-19 by molecular hydrogen therapy: Mechanisms and perspectives. *Oxidative Med. Cell. Longev.* **2021**, *2021*, 5513868. [CrossRef] [PubMed]
5. Ostojic, S.M. Molecular hydrogen in sports medicine: New therapeutic perspectives. *Int. J. Sports Med.* **2015**, *36*, 273–279. [CrossRef]
6. Wu, Q.; Su, N.; Cai, J.; Shen, Z.; Cui, J. Hydrogen-rich water enhances cadmium tolerance in Chinese cabbage by reducing cadmium uptake and increasing antioxidant capacities. *J. Plant Physiol.* **2015**, *175*, 174–182. [CrossRef]
7. Xu, S.; Zhu, S.; Jiang, Y.; Wang, N.; Wang, R.; Shen, W.; Yang, J. Hydrogen-rich water alleviates salt stress in rice during seed germination. *Plant Soil* **2013**, *370*, 47–57. [CrossRef]
8. Chen, M.; Cui, W.; Zhu, K.; Xie, Y.; Zhang, C.; Shen, W. Hydrogen-rich water alleviates aluminum-induced inhibition of root elongation in alfalfa via decreasing nitric oxide production. *J. Hazard. Mater.* **2014**, *267*, 40–47. [CrossRef]

9. Zeng, J.; Zhang, M.; Sun, X. Molecular hydrogen is involved in phytohormone signaling and stress responses in plants. *PLoS ONE* **2013**, *8*, e71038. [CrossRef]
10. Sun, Q.; Han, W.; Nakao, A. Biological safety of hydrogen. In *Hydrogen Molecular Biology and Medicine*; Springer: Dordrecht, The Netherlands, 2015; pp. 35–48.
11. Hylemon, P.B.; Harris, S.C.; Ridlon, J.M. Metabolism of hydrogen gases and bile acids in the gut microbiome. *FEBS Lett.* **2018**, *592*, 2070–2082. [CrossRef]
12. Ohsawa, I.; Ishikawa, M.; Takahashi, K.; Watanabe, M.; Nishimaki, K.; Yamagata, K.; Katsura, K.-I.; Katayama, Y.; Asoh, S.; Ohta, S. Hydrogen acts as a therapeutic antioxidant by selectively reducing cytotoxic oxygen radicals. *Nat. Med.* **2007**, *13*, 688–694. [CrossRef] [PubMed]
13. Winterbourn, C.C. The biological chemistry of hydrogen peroxide. *Methods Enzymol.* **2013**, *528*, 3–25. [PubMed]
14. Hess, D.T.; Stamler, J.S. Regulation by *S*-nitrosylation of protein post-translational modification. *J. Biol. Chem.* **2012**, *287*, 4411–4418. [CrossRef] [PubMed]
15. Shen, N.Y.; Bi, J.B.; Zhang, J.Y.; Zhang, S.M.; Gu, J.X.; Qu, K.; Liu, C. Hydrogen-rich water protects against inflammatory bowel disease in mice by inhibiting endoplasmic reticulum stress and promoting heme oxygenase-1 expression. *World J. Gastroenterol.* **2017**, *23*, 1375. [CrossRef] [PubMed]
16. Nguyen, T.K.; Lim, J.H. Is it a challenge to use molecular hydrogen for extending flower vase life? *Plants* **2022**, *11*, 1277. [CrossRef]
17. Kiaya, V. Post-harvest losses and strategies to reduce them. *Tech. Pap. Postharvest Losses Action Contre La Faim (ACF)* **2014**, *25*, 1–25.
18. Li, L.; Zeng, Y.; Cheng, X.; Shen, W. The applications of molecular hydrogen in horticulture. *Horticulturae* **2021**, *7*, 513. [CrossRef]
19. Fayaz, H.; Saidur, R.; Razali, N.; Anuar, F.S.; Saleman, A.R.; Islam, M.R. An overview of hydrogen as a vehicle fuel. *Renew. Sustain. Energy Rev.* **2012**, *16*, 5511–5528. [CrossRef]
20. Li, L.; Yin, Q.; Zhang, T.; Cheng, P.; Xu, S.; Shen, W. Hydrogen nanobubble water delays petal senescence and prolongs the vase life of cut carnation (*Dianthus caryophyllus* L.) flowers. *Plants* **2021**, *10*, 1662. [CrossRef]
21. Ali, S.; Nawaz, A.; Ejaz, S.; Haider, S.T.A.; Alam, M.W.; Javed, H.U. Effects of hydrogen sulfide on postharvest physiology of fruits and vegetables: An overview. *Sci. Hortic.* **2019**, *243*, 290–299. [CrossRef]
22. Guidotti, T.L. Hydrogen sulfide: Advances in understanding human toxicity. *Int. J. Toxicol.* **2010**, *29*, 569–581. [CrossRef] [PubMed]
23. Cheng, P.; Wang, J.; Zhao, Z.; Kong, L.; Lou, W.; Zhang, T.; Jing, D.; Yu, J.; Shu, Z.; Huang, L.; et al. Molecular hydrogen increases quantitative and qualitative traits of rice grain in field trials. *Plants* **2021**, *10*, 2331. [CrossRef] [PubMed]
24. Li, L.; Lou, W.; Kong, L.; Shen, W. Hydrogen commonly applicable from medicine to agriculture: From molecular mechanisms to the field. *Curr. Pharm. Des.* **2021**, *27*, 747–759. [CrossRef] [PubMed]
25. Zhao, Z.; Li, C.; Liu, H.; Yang, J.; Huang, P.; Liao, W. The involvement of glucose in hydrogen gas-medicated adventitious rooting in cucumber. *Plants* **2021**, *10*, 1937. [CrossRef]
26. Chen, H.; Zhang, J.; Hao, H.; Feng, Z.; Chen, M.; Wang, H.; Ye, M. Hydrogen-rich water increases postharvest quality by enhancing antioxidant capacity in *Hypsizygus marmoreus*. *AMB Express* **2017**, *7*, 1–10. [CrossRef]
27. Jiang, K.; Kuang, Y.; Feng, L.; Liu, Y.; Wang, S.; Du, H.; Shen, W. Molecular hydrogen maintains the storage quality of Chinese Chive through improving antioxidant capacity. *Plants* **2021**, *10*, 1095. [CrossRef]
28. Hu, H.; Li, P.; Wang, Y.; Gu, R. Hydrogen-rich water delays postharvest ripening and senescence of kiwifruit. *Food Chem.* **2014**, *156*, 100–109. [CrossRef]
29. Zhang, T.; Wang, Y.; Zhao, Z.; Xu, S.; Shen, W. Degradation of carbendazim by molecular hydrogen on leaf models. *Plants* **2022**, *11*, 621. [CrossRef]
30. Noctor, G.; Mhamdi, A.; Chaouch, S.; Han, Y.I.; Neukermans, J.; Marquez-Garcia, B.; Queval, G.; Foyer, C.H. Glutathione in plants: An integrated overview. *Plant Cell Environ.* **2012**, *35*, 454–484. [CrossRef]
31. Schwarzländer, M.; Fricker, M.D.; Müller, C.; Marty, L.; Brach, T.; Novak, J.; Sweetlove, L.J.; Hell, R.; Meyer, A.J. Confocal imaging of glutathione redox potential in living plant cells. *J. Microsc.* **2008**, *231*, 299–316. [CrossRef]
32. Garcia, P.C.; Rivero, R.M.; López-Lefebre, L.R.; Sánchez, E.; Ruiz, J.M.; Romero, L. Direct action of the biocide carbendazim on phenolic metabolism in tobacco plants. *J. Agric. Food Chem.* **2001**, *49*, 131–137. [CrossRef] [PubMed]
33. Zeng, J.; Yu, H. Integrated metabolomic and transcriptomic analyses to understand the effects of hydrogen water on the roots of *Ficus hirta Vahl*. *Plants* **2022**, *11*, 602. [CrossRef] [PubMed]
34. Kamimura, N.; Ichimiya, H.; Iuchi, K.; Ohta, S. Molecular hydrogen stimulates the gene expression of transcriptional coactivator PGC-1α to enhance fatty acid metabolism. *npj Aging Mech. Dis.* **2016**, *2*, 1–8. [CrossRef] [PubMed]
35. Hancock, J.T.; Russell, G. Downstream signalling from molecular hydrogen. *Plants* **2021**, *10*, 367. [CrossRef]
36. Penders, J.; Kissner, R.; Koppenol, W.H. ONOOH does not react with H_2: Potential beneficial effects of H_2 as an antioxidant by selective reaction with hydroxyl radicals and peroxynitrite. *Free Radic. Biol. Med.* **2014**, *75*, 191–194. [CrossRef]
37. Hancock, J.T.; LeBaron, T.W.; Russell, G. Molecular hydrogen: Redox reactions and possible biological interactions. *React. Oxyg. Species* **2021**, *11*, m17–m25. [CrossRef]
38. Peck, H.D. The ATP-dependent reduction of sulfate with hydrogen in extracts of *Desulfovibrio desulfuricans*. *Proc. Natl. Acad. Sci. USA* **1959**, *45*, 701–708. [CrossRef] [PubMed]
39. Hancock, J.T.; Hancock, T.H. Hydrogen gas, ROS metabolism, and cell signaling: Are hydrogen spin states important? *React. Oxyg. Species* **2018**, *6*, 389–395. [CrossRef]

40. Russell, G.; Zulfiqar, F.; Hancock, J.T. Hydrogenases and the role of molecular hydrogen in plants. *Plants* **2020**, *9*, 1136. [CrossRef]
41. Greening, C.; Islam, Z.F.; Bay, S.K. Hydrogen is a major lifeline for aerobic bacteria. *Trends Microbiol.* **2022**, *30*, 330–337. [CrossRef]

Review

Is It a Challenge to Use Molecular Hydrogen for Extending Flower Vase Life?

Toan Khac Nguyen and Jin Hee Lim *

Department of Plant Biotechnology, Sejong University, Seoul 05006, Korea; toanchrys@sju.ac.kr
* Correspondence: jinheelim@sejong.ac.kr

Abstract: Currently, molecular hydrogen treatment has the potential to manage the Corona Virus disease (COVID-19) and pandemic based on its anti-inflammatory, apoptosis-resistance, antioxidant, and hormone-regulating properties. Antioxidant properties are beneficial in both animal and human diseases. In agricultural sciences, molecular hydrogen is used to postpone postharvest ripening and senescence in fruits. However, studies on flower senescence are limited to the application of hydrogen molecules during floral preharvest and postharvest. Fortunately, improved tools involving molecular hydrogen can potentially improve postharvest products and storage. We also discuss the benefits and drawbacks of molecular hydrogen in floral preharvest and postharvest. This review provides an overview of molecular hydrogen solutions for floral preservative storage.

Keywords: cut flower; flower industry; postharvest quality; postharvest technique; the fourth industrial revolution; vase life

Citation: Nguyen, T.K.; Lim, J.H. Is It a Challenge to Use Molecular Hydrogen for Extending Flower Vase Life? . *Plants* **2022**, *11*, 1277. https://doi.org/10.3390/plants11101277

Academic Editor: John T. Hancock

Received: 6 April 2022
Accepted: 4 May 2022
Published: 10 May 2022

Publisher's Note: MDPI stays neutral with regard to jurisdictional claims in published maps and institutional affiliations.

1. Introduction

The floral market is defined by the high quality of the commercial standard pipeline from floral farmers to the final customers. The flowers, which are considered beautiful symbols of love, ceremony, appreciation, and respect, undergo discoloration, bending, or shriveling in commercialized sectors, reducing the quality of the floral products. Preharvest, vase life, and postharvest values are the most important characteristics for evaluating the commercial quality of cut flowers [1,2]. The preharvest quality of cut flowers is affected by temperature and seasonal variations [2]. Postharvest quality is influenced by multiple genetic factors, conditions of the preharvest environment [3], postharvest management practices throughout the supply chain [4], plant maturity, planting and harvesting seasons [5], nutritional status [6], water balance, and postharvest temperature [7]. Vase life and cut flower quality can be improved by evaluating appropriate genotypes in breeding programs [7–9], selecting the optimal temperature for storage and transport [10], and applying exogenous chemical substances such as sucrose [11], salicylic acid, glutamine [12], gibberellic acid [13], humic acid [14], and 1-methylcyclopropene [15].

To date, the COVID-19 pandemic has been restricted by vaccination and it can be potentially achieved by molecular hydrogen treatment through its characteristics, such as apoptosis-resistance, antioxidant, anti-inflammatory, and hormone-regulating properties. In agricultural sciences, molecular hydrogen is applied to postpone postharvest ripening and senescence in fruits. Hydrogen and its forms are well-known power energy carriers with numerous applications, and they can be easily transported. Hydrogen gas (H_2) has a broad range of biological effects and is a useful tool in medicine and agriculture [16–19]. H_2 affects plant growth, stress-tolerance enhancement [17], and possesses important effects on bacteria communities by preventing bacterial blockage and rot in xylem vessels [20]. H_2 has beneficial effects and improves the vase life quality of cut flowers, such as roses, by enhancing the beneficial bacteria abundances present on the stem-end cut surface [20]. The postharvest senescence of cut flowers results in significant commercial production losses, which are linked to a series of signaling molecules, such as magnesium hydride

(MgH$_2$) with H$_2$-releasing material [21], ethylene [22], ROS [23] (Figure 1), and nitric oxide (NO) [24]. Recently, the application of H$_2$ in the form of hydrogen-rich water (HRW) was shown to delay postharvest senescence and increase the quality of cut flowers [25–27]. H$_2$ can inhibit ethylene roles and affects signal transduction to regulate the expression of related genes (*Rh-ACS3*, *Rh-ACO1*, and *Rh-ETR1*), thus delaying flower senescence during the vase period [27]. In addition, H$_2$-stimulated NO can act as a downstream signaling molecule to maintain postharvest quality in cut lilies [28]. In this review, we discuss the benefits and drawbacks of molecular hydrogen in floral postharvest periods. This study considers the application of molecular hydrogen tools in floral preservation.

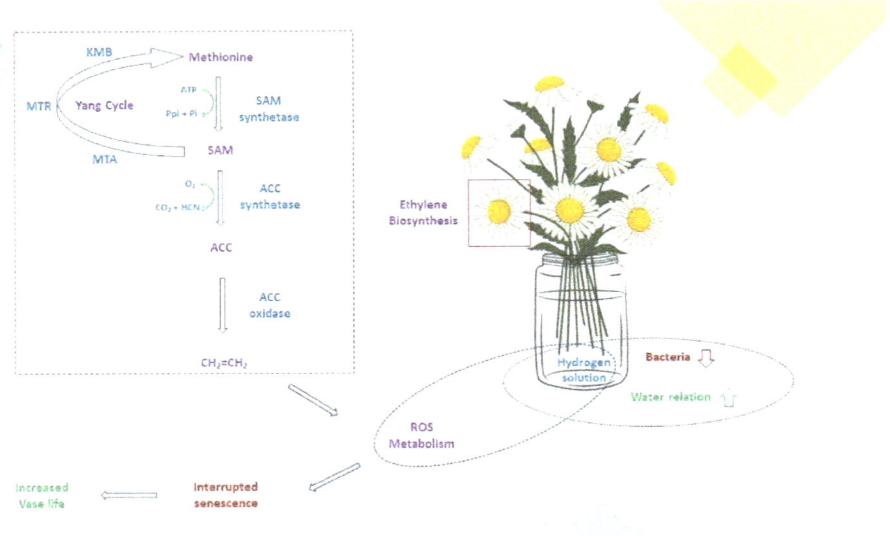

Figure 1. Possible roles of the effective hydrogen solution in floral preservative solution.

2. The Impact of Hydrogen Solution in Floral Preharvest and Postharvest

In roses, the quality of vase life is important for supporting innovative solutions that improve postharvest techniques [29]. In cut roses (*Rosa hybrida* 'Movie star'), the study of HRW showed a significantly extended vase life of cut roses by regulating the bacteria community of the stem ends [20]. HRW inhibited the bacterial blockages caused by bacteria colonization and biofilm formation in rose xylem vessels [20]. Therefore, it increased water uptake and extended cut rose vase life. By using high-throughput sequencing of the 16S rRNA gene sequence, it was concluded that HRW significantly developed the richness of bacterial communication on the stem-end cut surface [20]. The beneficial abundances were developed by 1% HRW on the stem-end cut surface, and it can be a key factor for prolonging flower vase life, especially in roses [20]. In another cut rose (*Rosa hybrida* 'Carola'), the study of H$_2$-releasing materials, such as MgH$_2$-treated cut rose flowers, is shown as an alternative tool for a more flexible and convenient hydrogen supply [21]. The effect of 0.001 g L^{-1} MgH$_2$-treated cut rose flowers was related to that of 10% HRW produced by electrolysis (similarly hereinafter) [21]. This study validated a critical role for the stimulated NO in the MgH$_2$-extended vase life of cut flowers [21].

In cut lily (*Lilium* spp.) flowers, treatment with HRW at 0.5% and 1% increased vase life and maintained maximum flower diameter [25]. In cut rose (*Rosa hybrid* L.) flowers, 50% HRW treatment significantly extended vase life and provided the maximum flower diameter [25]. The leaf relative water content and fresh weight of cut lilies and roses were improved by appropriate doses of HRW [25]. Compared with the control, the leaf stomata size was diminished in cut lily and rose flowers in the HRW treatment [25]. HRW treatment significantly decreased leaf MDA content, and reduced electrolyte leakage in cut lilies [25]. Both cut lily and rose flowers showed improved antioxidant enzyme activities [25]. Exogenously applied H_2 might increase vase life and improve postharvest quality in cut flowers by controlling water balance and membrane stability, and by reducing stomata size and oxidative damage [25].

In cut lilies (*Lilium* 'Manissa'), the relationship between H_2 and NO was studied, and differentially accumulated proteins were identified during postharvest freshness [28]. HRW (1%) and 150 µM sodium nitroprusside (SNP) significantly improved vase life and quality, whereas NO inhibitors suppressed the positive effects of HRW [28]. Proteomic analysis showed 50 differentially accumulated proteins in lily leaves, which were divided into seven functional categories [28]. Among them, ATP synthase CF1 alpha subunit (chloroplast) (AtpA) was up-regulated by HRW and down-regulated by the NO inhibitor [28]. NO might be affected by H_2-improved freshness of cut lilies, and the AtpA protein can play a critical role during this process [28].

Hydrogen nanobubble water (HNW) was used to screen cut carnation flowers (*Dianthus caryophyllus* L.) for delayed senescence [30]. Compared to conventional HRW, HNW had higher concentration properties and residence times for dissolved hydrogen gas [30]. The application of 5% HNW significantly increased the cut carnation vase life compared with distilled water, other doses of HNW (including 1%, 10%, and 50%), and 10% HRW, which aligned with the fresh weight and water content loss, provided electrolyte leakage, oxidative damage, and cell death in the petals [30]. The increasing trend in the activity of nucleases (including DNase and RNase) and proteases during vase life was prevented by 5% HNW [30]. Thus, HNW delayed petal senescence by reducing ROS accumulation and the initial activities of senescence-associated enzymes [30].

In daylily (*Hemerocallis fulva* L.) cultivar 'Dawuzui', HRW is used for preharvest treatment not only to increase bud yield, but also to maintain redox homeostasis by suppressing the gathering of $O_2^{\bullet-}$ and H_2O_2 in daylily buds under conditions of cold storage [31]. It prevents daylily bud sepal from browning during cold storage because it enhances membrane function, maintains fatty acid ratio, and reduces lipid peroxidation extension [31]. Moreover, the increasing total phenolics and the decreasing polyphenol oxidase activity also provide for the alleviation of bud browning [31].

In marigolds (*Tagetes erecta* L.), the use of 50% HRW showed physiological changes such as increasing root number and length of its explants [32]. Compared with the control, the use of hydrogen-rich water extended polyphenol oxidase, peroxidases, and indoleacetic acid oxidase activity [32]. Hydrogen gas promotes adventitious floral explant-root development by relatively increasing water content, metabolic constituents, rooting-related enzymes, and simultaneously maintaining cell membrane integrity (Table 1) [32].

Magnesium hydride (MgH_2), which is a suitable solid-state hydrogen source with high-capacity storage (7.6 wt%), was first applied as a hydrogen generation source with 98% purity and 0.5–25 µm size for floral postharvest preservation in cut carnation flowers [33]. Combining MgH_2 and citrate buffer solution could greatly increase efficiency compared to that of MgH_2 solutions in water [33]. The production and hydrogen residence time in solution were increased when compared with HRW [33]. Redox homeostasis was re-established and the progressing transcripts of representative senescence-associated genes, together with *DcbGal* and *DcGST1*, partly disappeared [33]. In contrast, the considered responses were blocked by the inhibition of endogenous H_2S with hypotaurine and H_2S collectors [33]. These results confirmed that MgH_2-supplying H_2 could extend cut carnation vase life via H_2S signaling, which could be a possible application of hydrogen-releasing methods in floral postharvest [33].

Table 1. An overview on the hydrogen forms used for floral treatments.

Hydrogen Forms	Floral Treatments	Utilization Treatment Parameters	Results	References
Hydrogen rich water (HRW)	Cut rose (*Rosa hybrida* 'Movie star')	1% HRW	Development of beneficial bacteria abundances on the stem-end cut surface.	[20]
	Daylily (*Hemerocallis fulva* L.) cultivar 'Dawuzui'	Preharvest: 0.8 μmol L^{-1} H$_2$	Improvement of yield and quality.	[31]
	Marigold (*Tagetes erecta* L.) explants	50% HRW	Induced root development	[32]
Hydrogen nanobubble water(HNW)	Carnation (*Dianthus caryophyllus* L.) cultivar 'Pink Diamond'	5% HNW	Development of the effective concentration and residence time of H$_2$ in water for extending vase life.	[30]
Magnesium hydride(MgH$_2$)	Carnation (*Dianthus caryophyllus* L.) cultivar 'Pink Diamond'	MgH$_2$ (0.1 g L^{-1}) with citrate	MgH$_2$-prolonged vase life of cut carnation flowers via increasing GST expression.	[33]
	Cut rose (*Rosa hybrida* 'Carola')	MgH$_2$ (0.001 g L^{-1}) with H$_2$-releasing donor	Re-establishing redox homeostasis to extend vase life	[21]

Endogenous ethylene production and ethylene gene expression in biosynthesis and signaling pathways were studied to determine the link between H$_2$ and ethylene during the senescence of cut roses [27]. The addition of exogenous ethylene to ethephon increased the senescence of cut roses, with 100 mg L^{-1} ethephon presenting the most obvious senescent phenotype [27]. The study of HRW (1%) indicated the best vase life quality by reducing ethylene production [27]. It decreased 1-aminocyclopropene-1-carboxylate (ACC) accumulation, as well as ACC synthase (ACS) and ACC oxidase (ACO) activities [27]. It also produced *Rh-ACS3* and *Rh-ACO1* expression in ethylene biosynthesis [27]. HRW increased the transcripts of ethylene receptor genes *Rh-ETR1* from day 4 to day 6 in the blooming period and suppressed *Rh-ETR3* at day 8 after harvest in the senescence phase [27]. The effect of HRW on *Rh-ETR1* and *Rh-ETR3* expression still existed when ethylene production was compromised by adequately adding exogenous ethylene in HRW-treated cut rose petals, and HRW directly repressed the protein level of *Rh-ETR3* in a transient expression assay [27].

3. The Potential Observation Using Hydrogen Tools in Floral Preservative Solution

Currently, the COVID-19 pandemic has impacted the global economy, including the flower industry. Thus, the preservative solution not only prolongs flower life but also prevents the substantial drop in prices of exporting flowers. Floral preservative solutions have been widely used by growers, florist sellers, and customers to extend vase life and maintain the quality of cut flowers [9]. Preservative solutions have many advantages, such as reducing bacterial agents in the vase, increasing water uptake, and balancing the carbohydrate requirement for metabolic cycle activities of cut flowers [9,34,35]. Floral preservative solutions can be separated into two types: chemical solutions and eco-friendly solutions [8]. Chemical preservative solutions, such as aluminum sulfate (Al$_2$(SO$_4$)$_3$), aminooxyacetic acid (AOA), benzyladenine (C$_{12}$H$_{11}$N$_5$), calcium, calcium nitrate, calcium dichloride (CaCl$_2$), chlorine compounds (sodium hypochlorite, sodium dichloroisocyanurate, chlorine dioxide (ClO$_2$), cobalt chloride (CoCl$_2$)), hydroquinone (HQ), 8-hydroxyquinoline sulfate (8-HQS), silver thiosulfate (STS), silver nitrate (AgNO$_3$), isothiazolinone, and quaternary ammonium chloride, can extend vase life, develop flower openings, and recover flower stem and size or petal color by balancing osmotic regulation [8,36]. Eco-friendly solutions can coincide with various factors such as prolonged vase life, controlled water uptake, and prevention of bacterial growth [8]. The H_2 solution was divided into an eco-friendly preservative. H_2 is known to affect cellular functions in plant cells [37]. HRW can extend the vase life of cut flowers, including carnations [30,38], roses [25], lisianthus [26], and lilies [28]. A minor drawback of H_2 in HRW is the residence time, which is commonly shorter than its present half-life in water of approximately 100 min [30]. However, H_2 application is advantageous in that it promotes the formation of nanobubbles with high internal pressure and negatively charged surfaces, which can increase the residence time and solubility in liquid [39].

HNW may have broad applications, not only in supporting human health care but also in extending the quality of floral life. HNW reduced ROS accumulation induced by senescence, thereby maintaining membrane integrity; HNW induces the initial inhibition of nuclease and protease activity, which may partially alleviate cell death, delay senescence, and prolong the life of flowers. In conclusion, molecular hydrogen can be applied to the floral industry for extending floral vase life, as long as the supplied tools of HNW-mediated H_2 show increasing availability of H_2, which has been a powerful tool in horticulture. Furthermore, they reduce ROS accumulation and inhibit the activities of proteases and nucleases.

Hydrogen is most frequently stored in tanks as gas or liquid for small-scale mobile and stationary applications. In general, geological storage is the best choice for large-scale and long-term storage, whereas tanks are more suitable for short-term and small-scale storage. The cost-benefit analysis of H_2 application in floral preservatives postharvest does not sufficiently compare chemical and eco-friendly solutions. Although renewable H_2 is expensive, innovative technologies, such as water electrolysis, are estimated to reduce production costs. Thus, the estimated cost of H_2 application is mainly dependent on labor costs under economic conditions [40]. However, chemical effects in physiological situations have not been established. There are various ways to regulate the senescence of cut flowers, such as NO, calcium ion (Ca^{2+})/calmodulin (CaM) [41], sodium hypochlorite + aminoisobutyric acid + 1-methylcyclopropene (ClAM) [29], and sucrose + ClO_2 [42]. When Ca^{2+} chelators, Ca^{2+} channel inhibitors, and CaM antagonists are applied, the promoting effects of NO on vase life are blocked [41]. The Ca^{2+} channel inhibitor nifedipine itself negatively impacts fresh-keeping by inhibiting endogenous Ca^{2+} [41]. Hydrogen solution can be preferred over other methods [40,43]. Hydrogen solution is active against a broad range of micro-organisms, including bacteria, yeasts, and fungi, and is eco-friendly [40,43]. We expect that in the period of low-carbon agriculture, H_2 presents unique renewable and eco-friendly solutions for the environment and people, while also reducing greenhouse gas emissions on ignition.

4. Further Prospects for Hydrogen Treatment in the Floral Industry in Korea

In Korea, the Korea Seed & Variety Service noted that, of the 7731 crops filed and registered to date, flowers constituted 4123 representing 53% of the total registered crops [44]. Since the 1980s, floral genetic resources have been focused on culturing experiments with floral varieties such as chrysanthemum, rose, trumpet lily, and carnation [44]. In the following ten years, global agricultural products have enabled the introduction of new flower varieties and seedlings for export [44]. In 1995, Korea joined the International Union of the Production of New Varieties of Plants (UPOV), which included various studies on breeding and high-quality seedlings of chrysanthemums, roses, lilies, carnations, hibiscus, and gerberas [44]. During the 2000s, breeding technology was stabilized leading to many new varieties, increasing the ingression rates of chrysanthemums, orchids, and roses from 1% in 2000 to 5.8% in 2008, and 27.3% in the 2010s [44]. In Korea, there are some representative domestic varieties of breeding samples such as "Baekma" (chrysanthemums), "Deep purple" (rose), "Woori tower" (lily), and "Shiny gold" (freesia) [44].

For the Korean floriculture industry, it could be beneficial to use HRW and HNW, which are cheap, eco-friendly, non-toxic to humans, and provide a long life for cut flowers. H_2 can be linked to plant stresses, such as temperature, heavy metals, salinity, and light stress, which is promising for the use of H_2 treatment to delay postharvest senescence

(Table 2) [44]. However, the effects of HRW are visible during postharvest if the plants are also treated at preharvest [45]. HRW has a short residence time with a half-life of approximately 100 min in water [30]. HNW diminishes ROS accumulation and is associated with delayed response senescence and extended flower vase life. H_2 is approved by other industries [46], which are similar in that its creation, storage, and transport costs will become cheaper, combined with an attractive sense in agricultural production [40,44]. H_2 treatments, which are representative solutions such as HRW or HNW, could be associated with other treatments, including fertilizers, also resulting in lower costs in the floral industry. Even if the current costs are excessive, the application of H_2-based treatments is likely to be efficient in the future, and these may be extremely promising for a range of postharvest uses [40,44]. Although few studies exist on using hydrogen treatment in the floral industry, H_2 can be used in solution or donor molecule forms and can improve the quality of floral postharvest. In postharvest solutions, especially in the floral industry, the use of many H_2-based treatments is expected to investigate the optimization of H_2 delivery methods and provide solutions that are suitable to the crop being used. This is a safe, eco-friendly, and easy way of using H_2 and its form for application in the floral postharvest and horticultural industries in Korea and internationally. Further investigation of H_2-based treatments in Korea could expand, as could the development of innovative tools, which would be re-affirmed by cost-benefits analysis.

Table 2. Collection of H_2 treatments (Hydrogen-rich water—HRW, Hydrogen nanobubble water—HNW) for cut flowers in postharvest. H_2 concentrations are converted from the information given in the reference papers [43].

Flower Investigation	Treatment	Result	Reference
Rose 'Movie star'	1% HRW (best in 0.00235 mM H_2)	Less flower senescence. Investigated by ethylene metabolism.	[27]
Lily (*Lilium* spp.) and rose (*Rosa hybrid* L.)	Lily: 0.5% HRW (2.25 µM H_2) and 1% (4.5 µM H_2); Rose: 50% HRW (0.225 mM H_2)	Extended vase life. Greater flower diameter. Reduced oxidative stress.	[25]
Lily (*Lilium* 'Manissa')	1% HRW (0.0022 mM H_2) and 150 µM sodium nitroprusside (SNP)	Improved flower freshness. ATP synthase CF1 alpha subunit (AtpA) up-regulated.	[28]
Lisianthus (*Eustoma grandiflorum*)	HRW (0.078 mM H_2)	Vase life prolonged. Redox maintained as reducing oxidative stress.	[26]
Carnation (*Dianthus caryophyllus* L.).	Hydrogen nanobubble water (5% HNW): best in 0.025 mM H_2	Less senescence leading to prolonged vase life. Minimized oxidative stress.	[30]

5. Conclusions

This paper considered eco-friendly tools to improve cut flower vase life and is intended not only to help scientists, especially florists, to understand hydrogen technologies but also to provide an overview of steps for keeping cut flowers with a long vase life. The use of hydrogen solutions for cut flowers must be investigated and developed (Figure 2), and innovative tools should be provided based on their suitability for the environment and human health. These hydrogen-based treatments should be considered and investigated for their benefits related to Korean floral postharvest.

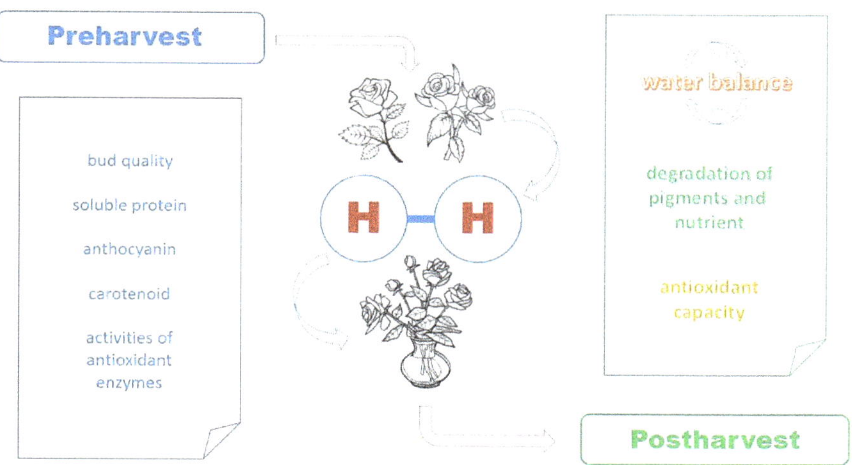

Figure 2. The physiological diagram for the study of floral senescence causes and hydrogen solution.

Author Contributions: T.K.N. wrote and revised the manuscript. J.H.L. designed and supervised the project. All authors have read and agreed to the published version of the manuscript.

Funding: This work was supported by the faculty research fund of Sejong University in 2022.

Institutional Review Board Statement: Not applicable.

Informed Consent Statement: Not applicable.

Data Availability Statement: Not applicable.

Acknowledgments: We would like to thank all members of the Floriculture Lab.

Conflicts of Interest: The authors declare no conflict of interest regarding the publication of this paper.

References

1. Liao, W.-B.; Zhang, M.-L.; Yu, J.-H. Role of nitric oxide in delaying senescence of cut rose flowers and its interaction with ethylene. *Sci. Hortic.* **2013**, *155*, 30–38. [CrossRef]
2. In, B.-C.; Seo, J.Y.; Lim, J.H. Preharvest environmental conditions affect the vase life of winter-cut roses grown under different commercial greenhouses. *Hortic. Environ. Biotechnol.* **2016**, *57*, 27–37. [CrossRef]
3. Onozaki, T.; Ikeda, H.; Yamaguchi, T. Genetic improvement of vase life of carnation flowers by crossing and selection. *Sci. Hortic.* **2001**, *87*, 107–120. [CrossRef]
4. Dolan, C.; Humphrey, J. Governance and trade in fresh vegetables: The impact of UK supermarkets on the African horticulture industry. *J. Dev. Stud.* **2000**, *37*, 147–176. [CrossRef]
5. Pompodakis, N.; Terry, L.; Joyce, D.; Lydakis, D.; Papadimitriou, M. Effect of seasonal variation and storage temperature on leaf chlorophyll fluorescence and vase of cut roses. *Postharvest Biol. Technol.* **2005**, *36*, 1–8. [CrossRef]
6. Azad, A.K.; Ishikawa, T.; Ishikawa, T.; Sawa, Y.; Shibata, H. Intracellular energy depletion triggers programmed cell death during petal senescence in tulip. *J. Exp. Bot.* **2008**, *59*, 2085–2095. [CrossRef]
7. Fanourakis, D.; Pieruschka, R.; Savvides, A.; Macnish, A.J.; Sarlikioti, V.; Woltering, E.J. Sources of vase life variation in cut roses: A review. *Postharvest Biol. Technol.* **2013**, *78*, 1–15. [CrossRef]
8. Nguyen, T.K.; Lim, J.H. Do eco-friendly floral preservative solutions prolong vase life better than chemical solutions? *Horticulturae* **2021**, *7*, 415. [CrossRef]
9. Nguyen, T.K.; Jung, Y.O.; Lim, J.H. Tools for cut flower for export: Is it a genuine challenge from growers to customers? *Flower Res. J.* **2020**, *28*, 241–249. [CrossRef]
10. Mohd Rafdi, H.; Joyce, D.; Lisle, A.; Li, X.; Irving, D.; Gupta, M. A retrospective study of vase life determinants for cut Acacia holosericea foliage. *Sci. Hortic.* **2014**, *180*, 254–261. [CrossRef]
11. Arrom, L.; Munné-Bosch, S. Sucrose accelerates flower opening and delays senescence through a hormonal effect in cut lily flowers. *Plant Sci.* **2012**, *188–189*, 41–47. [CrossRef] [PubMed]

12. Zamani, S.; Kazemi, M.; Aran, M. Postharvest life of cut rose flowers as affected by salicylic acid and glutamin. *World Appl. Sci. J.* **2011**, *12*, 1621–1624.
13. Saeed, T.; Hassan, I.; Abbasi, N. Effect of gibberellic acid on the vase life and oxidative activities in senescing cut gladiolus flowers. *Plant Growth Regul.* **2014**, *72*, 89–95. [CrossRef]
14. Fan, H.-m.; Li, T.; Sun, X.; Sun, X.-z.; Zheng, C.-s. Effects of humic acid derived from sediments on the postharvest vase life extension in cut chrysanthemum flowers. *Postharvest Biol. Technol.* **2015**, *101*, 82–87. [CrossRef]
15. Nergi, M.; Ahmadi, N. Effects of 1-MCP and ethylene on postharvest quality and expression of senescence-associated genes in cut rose cv. Sparkle. *Sci. Hortic.* **2014**, *166*, 78–83. [CrossRef]
16. Ohsawa, I.; Ishikawa, M.; Takahashi, K.; Watanabe, M.; Nishimaki, K.; Yamagata, K.; Katsura, K.-i.; Katayama, Y.; Asoh, S.; Ohta, S. Hydrogen acts as a therapeutic antioxidant by selectively reducing cytotoxic oxygen radicals. *Nat. Med.* **2007**, *13*, 688–694. [CrossRef]
17. Xie, Y.; Mao, Y.; Lai, D.; Zhang, W.; Shen, W. H2 Enhances Arabidopsis salt tolerance by manipulating ZAT10/12-mediated antioxidant defence and controlling sodium exclusion. *PLoS ONE* **2012**, *7*, e49800. [CrossRef]
18. Zeng, J.; Zhang, M.; Sun, X. Molecular hydrogen is involved in phytohormone signaling and stress responses in plants. *PLoS ONE* **2013**, *8*, e71038. [CrossRef]
19. Wu, Y.; Yuan, M.; Song, J.; Chen, X.; Yang, H. Hydrogen gas from inflammation treatment to cancer therapy. *ACS Nano* **2019**, *13*, 8505–8511. [CrossRef]
20. Fang, H.; Wang, C.; Wang, S.; Liao, W. Hydrogen gas increases the vase life of cut rose 'Movie star' by regulating bacterial community in the stem ends. *Postharvest Biol. Technol.* **2021**, *181*, 111685. [CrossRef]
21. Li, Y.; Li, L.; Wang, S.; Liu, Y.; Zou, J.; Ding, W.; Du, H.; Shen, W. Magnesium hydride acts as a convenient hydrogen supply to prolong the vase life of cut roses by modulating nitric oxide synthesis. *Postharvest Biol. Technol.* **2021**, *177*, 111526. [CrossRef]
22. Kumar, N.; Srivastava, G.C.; Dixit, K. Senescence in rose (*Rosa hybrida* L.): Role of the endogenous anti-oxidant system. *J. Hortic. Sci. Technol.* **2008**, *83*, 125–131. [CrossRef]
23. van Doorn, W.G.; Woltering, E.J. Physiology and molecular biology of petal senescence. *J. Exp. Bot.* **2008**, *59*, 453–480. [CrossRef] [PubMed]
24. Naing, A.H.; Lee, K.; Arun, M.; Lim, K.B.; Kim, C.K. Characterization of the role of sodium nitroprusside (SNP) involved in long vase life of different carnation cultivars. *BMC Plant Biol.* **2017**, *17*, 149. [CrossRef]
25. Ren, P.-J.; Jin, X.; Liao, W.-B.; Wang, M.; Niu, L.-J.; Li, X.-P.; Xu, X.-T.; Zhu, Y.-C. Effect of hydrogen-rich water on vase life and quality in cut lily and rose flowers. *Hortic. Environ. Biotechnol.* **2017**, *58*, 576–584. [CrossRef]
26. Su, J.; Nie, Y.; Zhao, G.; Cheng, D.; Wang, R.; Chen, J.; Zhang, S.; Shen, W. Endogenous hydrogen gas delays petal senescence and extends the vase life of lisianthus cut flowers. *Postharvest Biol. Technol.* **2019**, *147*, 148–155. [CrossRef]
27. Wang, C.; Fang, H.; Gong, T.; Zhang, J.; Niu, L.; Huang, D.; Huo, J.; Liao, W. Hydrogen gas alleviates postharvest senescence of cut rose 'Movie star' by antagonizing ethylene. *Plant Mol. Biol.* **2020**, *102*, 271–285. [CrossRef]
28. Huo, J.; Huang, D.; Zhang, J.; Fang, H.; Wang, B.; Wang, C.; Ma, Z.; Liao, W. Comparative proteomic analysis during the involvement of nitric oxide in hydrogen gas-improved postharvest freshness in cut lilies. *Int. J. Mol. Sci.* **2018**, *19*, 3955. [CrossRef]
29. Ha, S.T.T.; Lim, J.H.; In, B.-C. Extension of the vase life of cut roses by both improving water relations and repressing ethylene responses. *J. Hortic. Sci. Technol.* **2019**, *37*, 65–77. [CrossRef]
30. Li, L.; Yin, Q.; Zhang, T.; Cheng, P.; Xu, S.; Shen, W. Hydrogen nanobubble water delays petal senescence and prolongs the vase life of cut carnation (*Dianthus caryophyllus* L.) flowers. *Plants* **2021**, *10*, 1662. [CrossRef]
31. Hu, H.; Li, P.; Shen, W. Preharvest application of hydrogen-rich water not only affects daylily bud yield but also contributes to the alleviation of bud browning. *Sci. Hortic.* **2021**, *287*, 110267. [CrossRef]
32. Zhu, Y.; Liao, W. The metabolic constituent and rooting-related enzymes responses of marigold explants to hydrogen gas during adventitious root development. *Theor. Exp. Plant Physiol.* **2017**, *29*, 77–85. [CrossRef]
33. Li, L.; Liu, Y.; Wang, S.; Zou, J.; Ding, W.; Shen, W. Magnesium hydride-mediated sustainable hydrogen supply prolongs the vase life of cut carnation flowers via hydrogen sulfide. *Front. Plant Sci.* **2020**, *11*, 595376. [CrossRef] [PubMed]
34. Ha, S.T.T.; Kim, Y.-T.; Jeon, Y.H.; Choi, H.W.; In, B.-C. Regulation of *Botrytis cinerea* infection and gene expression in cut roses by using nano silver and salicylic acid. *Plants* **2021**, *10*, 1241. [CrossRef] [PubMed]
35. Ha, S.T.T.; Jung, Y.-O.; Lim, J.-H. Pretreatment with *Scutellaria baicalensis* Georgi extract improves the postharvest quality of cut roses (*Rosa hybrida* L.). *Hortic. Environ. Biotechnol.* **2020**, *61*, 511–524. [CrossRef]
36. Reid, M. Handling of Cut Flowers for Export. 2009. Available online: https://ucanr.edu/sites/Postharvest_Technology_Center_/files/231308.pdf (accessed on 3 May 2022).
37. Hancock, J.T.; Russell, G. Downstream signaling from molecular hydrogen. *Plants* **2021**, *10*, 367. [CrossRef]
38. Cai, M.; Du, H.-m. Effects of hydrogen-rich water pretreatment on vase life of carnation (*Dianthus caryophyllus*) cut flowers. *J. Shanghai Jiaotong Univ. (Agric. Sci.)* **2015**, *33*, 41–45.
39. Temesgen, T.; Bui, T.T.; Han, M.; Kim, T.-i.; Park, H. Micro and nanobubble technologies as a new horizon for water-treatment techniques: A review. *Adv. Colloid Interface Sci.* **2017**, *246*, 40–51. [CrossRef]
40. Li, L.; Zeng, Y.; Cheng, X.; Shen, W. The applications of molecular hydrogen in horticulture. *Horticulturae* **2021**, *7*, 513. [CrossRef]
41. Zhang, J.; Fang, H.; Huo, J.; Huang, D.; Wang, B.; Liao, W. Involvement of calcium and calmodulin in nitric oxide-regulated senescence of cut lily flowers. *Front. Plant Sci.* **2018**, *9*, 1284. [CrossRef]

42. Lee, Y.E.; Kim, W.S. Improving vase life and keeping quality of cut rose flowers using a chlorine dioxide and sucrose holding solution. *Hortic. Sci. Technol.* **2018**, *36*, 380–387. [CrossRef]
43. Hancock, J.T.; LeBaron, T.W.; May, J.; Thomas, A.; Russell, G. Molecular hydrogen: Is this a viable new treatment for plants in the UK? *Plants* **2021**, *10*, 2270. [CrossRef] [PubMed]
44. Kim, W.H.; Park, P.H.; Jung, J.A.; Park, K.Y.; Suh, J.-N.; Kwon, O.K.; Yoo, B.S.; Lee, S.Y.; Park, P.M.; Choi, Y.J.; et al. Achievement of flower breeding in Korea and its prospects. *Korean J. Breed. Sci.* **2020**, *52*, 161–169. (In Korean) [CrossRef]
45. Hu, H.; Zhao, S.; Li, P.; Shen, W. Hydrogen gas prolongs the shelf life of kiwifruit by decreasing ethylene biosynthesis. *Postharvest Biol. Technol.* **2018**, *135*, 123–130. [CrossRef]
46. Singh, S.; Jain, S.; Ps, V.; Tiwari, A.K.; Nouni, M.R.; Pandey, J.K.; Goel, S. Hydrogen: A sustainable fuel for future of the transport sector. *Renew. Sustain. Energy Rev.* **2015**, *51*, 623–633. [CrossRef]

Article

Hydrogen Nanobubble Water Delays Petal Senescence and Prolongs the Vase Life of Cut Carnation (*Dianthus caryophyllus* L.) Flowers

Longna Li [1], Qianlan Yin [1], Tong Zhang [1], Pengfei Cheng [1], Sheng Xu [2] and Wenbiao Shen [1,3,*]

[1] Laboratory Center of Life Sciences, College of Life Sciences, Nanjing Agricultural University, Nanjing 210095, China; lln2013034@njau.edu.cn (L.L.); 10117118@njau.edu.cn (Q.Y.); 2020116098@stu.njau.edu.cn (T.Z.); 2020216037@njau.edu.cn (P.C.)

[2] Institute of Botany, Jiangsu Province and Chinese Academy of Sciences, Nanjing 210014, China; xusheng@cnbg.net

[3] Center of Hydrogen Science, Shanghai Jiao Tong University, Shanghai 200240, China

* Correspondence: wbshenh@njau.edu.cn; Tel.: +86-25-84-399-032; Fax: +86-25-84-396-542

Abstract: The short vase life of cut flowers limits their commercial value. To ameliorate this practical problem, this study investigated the effect of hydrogen nanobubble water (HNW) on delaying senescence of cut carnation flowers (*Dianthus caryophyllus* L.). It was observed that HNW had properties of higher concentration and residence time for the dissolved hydrogen gas in comparison with conventional hydrogen-rich water (HRW). Meanwhile, application of 5% HNW significantly prolonged the vase life of cut carnation flowers compared with distilled water, other doses of HNW (including 1%, 10%, and 50%), and 10% HRW, which corresponded with the alleviation of fresh weight and water content loss, increased electrolyte leakage, oxidative damage, and cell death in petals. Further study showed that the increasing trend with respect to the activities of nucleases (including DNase and RNase) and protease during vase life period was inhibited by 5% HNW. The results indicated that HNW delayed petal senescence of cut carnation flowers through reducing reactive oxygen species accumulation and initial activities of senescence-associated enzymes. These findings may provide a basic framework for the application of HNW for postharvest preservation of agricultural products.

Keywords: hydrogen nanobubble water; vase life; senescence-associated enzymes; cut carnation flowers

Citation: Li, L.; Yin, Q.; Zhang, T.; Cheng, P.; Xu, S.; Shen, W. Hydrogen Nanobubble Water Delays Petal Senescence and Prolongs the Vase Life of Cut Carnation (*Dianthus caryophyllus* L.) Flowers. *Plants* **2021**, *10*, 1662. https://doi.org/10.3390/plants10081662

Academic Editors: John T. Hancock and Joshua D. Klein

Received: 8 July 2021
Accepted: 11 August 2021
Published: 12 August 2021

Publisher's Note: MDPI stays neutral with regard to jurisdictional claims in published maps and institutional affiliations.

1. Introduction

The rapid senescence of cut flowers during postharvest periods limits their economic value. Flower senescence is a coordinated and complex process, which is primarily related to loss of water, leakage of ions, generation of reactive oxygen species (ROS), and synthesis and degradation of proteins and nucleic acids [1]. ROS are involved in membrane degradation and contribute to the cell death. During the postharvest period for cut flowers, ROS overproduction was commonly observed, while scavenging of ROS may delay the onset of cut flower senescence as a result of increasing activities of antioxidant enzymes [1–3].

In the petals of *Petunia* flowers, activities of five nucleases increased [4], and RNase activity in petals of *Hemerocallis* did so during senescence [5]. Additionally, the degradation of protein exhibits a crucial role in the flower senescence, commonly accompanied with increased protease activity [1]. Accordingly, chemical inhibition of protease delayed the time to visible senescence in *Sandersonia* [6] and *Iris* [7] flowers.

Hydrogen gas (H_2), considered to be a selective antioxidant, has so far been primarily used in medicine [8,9]. Interestingly, the production of H_2 has been observed in plants under the normal or stressed conditions [10–12], although its detailed synthetic pathway (s) are not fully elucidated. Further studies have revealed that H_2 played an important role in defense responses of plants to abiotic stresses [13–15], plant growth [16], and secondary metabolism [17]. H_2 was beneficial to postharvest preservation of fruits (kiwifruit [18,19] and tomato [20]) and cut flowers (rose [21], lily [22], and *Lisianthus* [23]). Previous studies have reported that H_2, dissolved in water [24] or supplied by a H_2-releasing material (magnesium hydride [MgH_2]) [25,26], can prolong the vase life of cut carnation (*Dianthus Caryophyllus* L.) flowers by enhancing activities of antioxidant enzymes and by the involvement of other gaseous signaling molecules (including nitric oxide and hydrogen sulfide).

At present, hydrogen-rich water (HRW) is a main source for exogenous H_2 delivery, while H_2 originates mainly from electrolytic water or gas cylinder [27]. However, the low solubility and short residence time of dissolved H_2 limits its wide-spread application. Fortunately, the nanobubble technology provides an opportunity to overcome these disadvantages. Nanobubbles (less than 500 nm in diameter) have several properties including large surface area, high internal pressure, and negatively charged surface (zeta potential), which accelerate dissolution of the gas into the liquids and remain its stability in the liquids for longer times [28]. Hydrogen nanobubble water (HNW) was observed to exhibit higher antioxidant activity than conventional HRW without nanobubbles [29]. Previous results showed that HNW improved copper tolerance in *Daphnia magna* by reducing oxidative damage [30]. Although solid H_2-storage materials (such as MgH_2) can also improve the solubility and residence time of H_2 in water, their potential threat to the environment should be concerned, especially when they are extensively used in agriculture. Thus, without conventional chemical additives except H_2, HNW is more environmentally friendly.

In this study, we aimed to identify the benefits of HNW in prolonging the vase life of cut carnation flowers. It was confirmed that HNW was a superior source to delay cut flower senescence in comparison with HRW. Further experiments showed that HNW reduced ROS accumulation and the initial activities of DNase, RNase, and protease. These findings provide a basic idea for application of HNW in postharvest preservation of agricultural products.

2. Results

2.1. Effects of HNW on the Vase Life of Cut Carnation Flowers

As shown in Figure 1, the initial content of H_2 in fresh HNW (about 1.0 µg mL^{-1}; regarded as 100% saturation HNW) was higher than that in ordinary fresh HRW without nanobubbles (about 0.8 µg mL^{-1}; also regarded as 100% saturation HRW). Meanwhile, H_2 remained in HNW for about 6 h, which was longer than that in HRW (about 4 h) as the result of the slower evolution. Afterwards, 100% HNW or 100% HRW was immediately diluted to the required concentration (1%, 5%, 10%, 50%, or 10% (*v/v*); equivalently as about 0.01, 0.05, 0.1, 0.5, or 0.08 µg H_2 mL^{-1}), respectively.

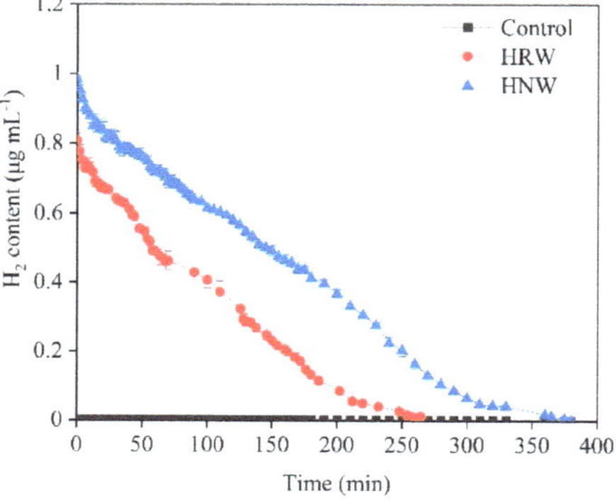

Figure 1. Changes in H_2 content of fresh hydrogen nanobubble water (HNW) and hydrogen-rich water (HRW).

Cut carnation flowers incubated in distilled water (control) and HNW/HRW with different concentrations were photographed to document the symptoms of senescence. Compared with distilled water, 1%, 5%, 10% HNW, or 10% HRW differentially delayed the petal wilting and flower withering (Figure 2A), while no significant response was observed for 50% HNW. The treatments with 1%, 5% HNW, and 10% HRW significantly prolonged the vase life of cut carnation flowers, which were assessed as 8.7 ± 0.7 d, 10.6 ± 1.0 d, or 8.6 ± 0.4 d, that prolonged vase life by 23.9%, 50.9%, or 22.3%, respectively, over the H_2-free control (7.0 ± 0.5 d; Figure 2B). Notably, 5% HNW displayed the most obvious positive effect, which was also greater than 10% HRW.

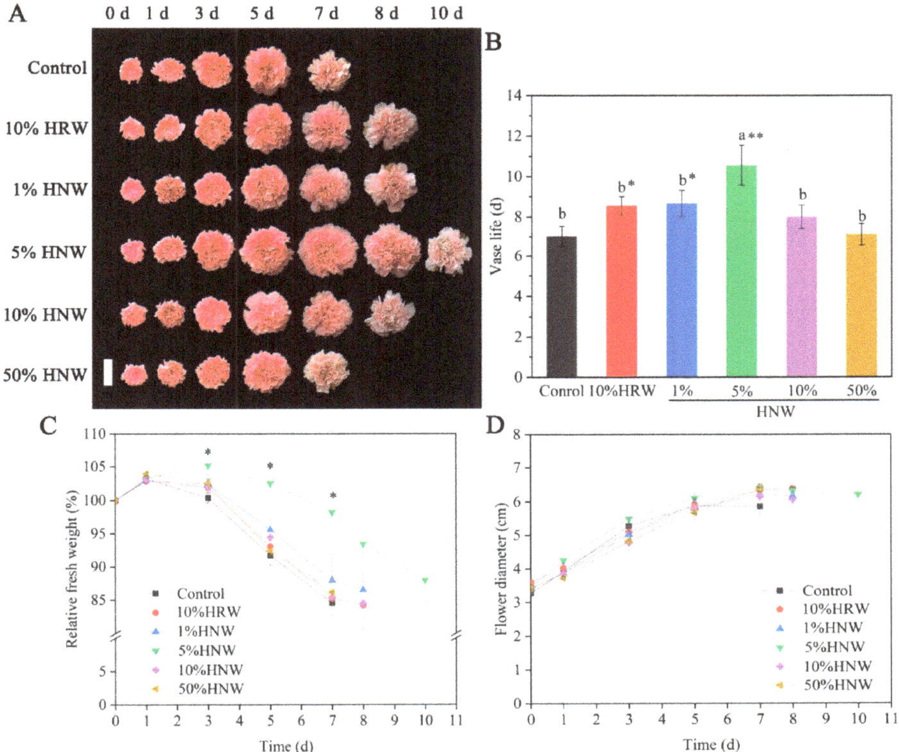

Figure 2. Effects of different concentrations hydrogen nanobubble water (HNW) and 10% hydrogen-rich water (HRW) on morphological changes (**A**), vase life (**B**), relative fresh weight (RFW; (**C**)), and flower diameter (**D**) of cut carnation flowers. Cut flowers were incubated in distilled water (control), 1%, 5%, 10%, and 50% HNW, and 10% HRW for 3 d (changed daily), and then, in distilled water, which was replaced daily until the end of experiments. Representative photographs of cut flowers were taken (scale bar = 4 cm). Afterward, vase life (**B**), RFW (**C**), and flower diameter (**D**) were expressed as mean ± standard error (SE). There were three replicates and three flowers per each. Experiments were conducted three times. The different letters in a column indicated significant differences according to Duncan's multiple range test ($p < 0.05$), and asterisks indicated significant differences in comparison with the control at $p < 0.05$ (*) or 0.01 (**) according to *t*-test.

During vase life period, the relative fresh weight (RFW) of cut carnation flowers initially increased and then decreased (Figure 2C). Five percent HNW remarkably extended the RFW increase for 3 days, then postponed and slowed down the weight loss on the following days, in comparison with other treatments. However, compared with the distilled water, HNW or HRW had no significant effect on flower diameters (Figure 2D). Since 5% HNW treatment displayed the optimal vase life response, it was used for the subsequent experiments.

2.2. Effects of HNW on Relative Water Content and Electrolyte Leakage of Cut Carnation Flowers

In this experiment, the relative water content (RWC) of cut carnation flowers treated with distilled water was continuously decreased during senescence period (Figure 3A). Comparatively, RWC of carnation petals treated with 5% HNW was significantly higher than that of the control on day 3, while thereafter, it was not obviously different from control.

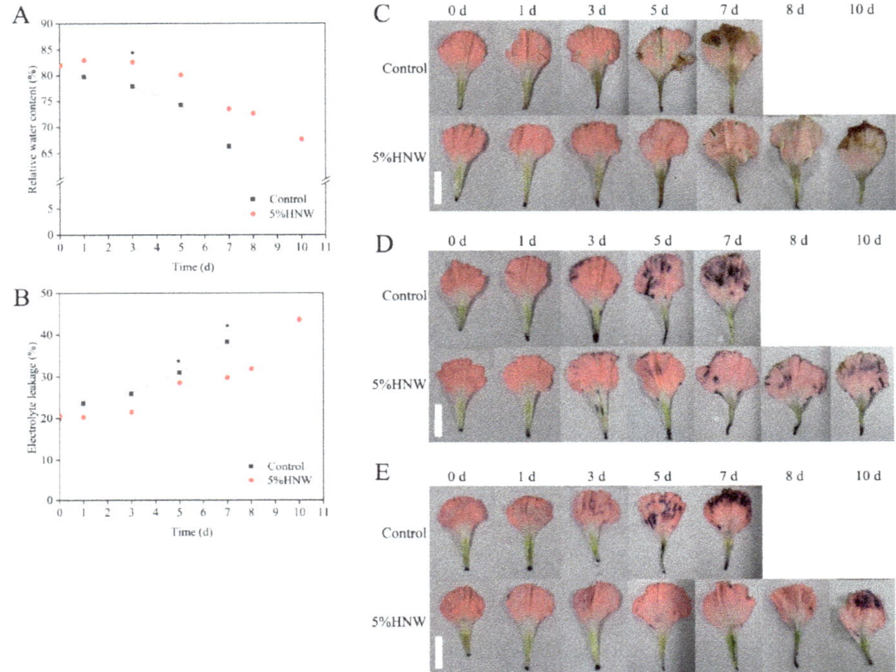

Figure 3. Changes in relative water content (RWC; (**A**)), electrolyte leakage (**B**), hydrogen peroxide (H_2O_2; (**C**)), and superoxide anions (O_2^-; (**D**)) accumulation as well as cell death (**E**) of carnation petals during vase period. Cut flowers were incubated in distilled water (control) and 5% HNW, respectively, for 3 d (changed daily), and then, in distilled water, which was replaced daily until the end of experiments. Values were expressed as the means ± SE. Asterisks indicated significant differences at $p < 0.05$ by *t*-test. The petals were stained with 3,3-diaminobenzidine (DAB; (**C**)), nitro blue tetrazolium (NBT; (**D**)), and trypan blue (**E**), respectively, then photographed with digital camera. Scale bar = 4 cm. Three flowers per replicate were selected, and total flowers in triplicate were 9 (3 × 3) for each treatment at each time point, then photographed.

The changes in electrolyte leakage in petals are supposed to indicate changes in membrane permeability [31]. Electrolyte leakage values of carnation petals increased continuously during senescence period (Figure 3B), confirming that the integrity of cell membrane was gradually impaired. While, 5% HNW postponed the increase in electrolyte leakage and maintained it in lower levels at 5–7 d of vase life, in comparison with the control.

2.3. Effects of HNW on Reducing the Oxidative Damage and Cell Death in Carnation Petals

3,3-Diaminobenzidine (DAB) and nitro blue tetrazolium (NBT) staining was usually used to detect the accumulation of ROS (hydrogen peroxide (H_2O_2) and superoxide anions (O_2^-). A gradual increase in DAB- and NBT-dependent staining in the control during vase period was observed, respectively (Figure 3C,D), suggesting continuous accumulation of ROS and disrupted cellular redox homeostasis. Comparatively, petals from flowers treated with 5% HNW had a slight staining.

In addition, senescence can cause cells death, which was commonly detected by trypan blue staining. As expected, results of trypan blue staining showed that petals treated with 5% HNW exhibited slight blue coloration (Figure 3E). These staining results indicated that the accumulation of ROS and cell death of petals were delayed by 5% HNW.

2.4. Effects of HNW on the Activities of DNase, RNase, and Protease

To further investigate the contribution of HNW, changes in the enzymatic activities of nucleases (including DNase and RNase) and protease were determined. These assays showed that the activities of DNase and RNase gradually increased during senescence in the control sample (Figure 4A,B). By contrast, when treated with 5% HNW, the activities of DNase and RNase initially decreased and remained at a low level up to day 7, then increased during the last 3 days. Additionally, protease activity in the control petals initially increased (3 d) and, thereafter, declined to some extent (Figure 4C). Compared with the control, the peak of protease activity was postponed to the 7th day by 5% HNW treatment.

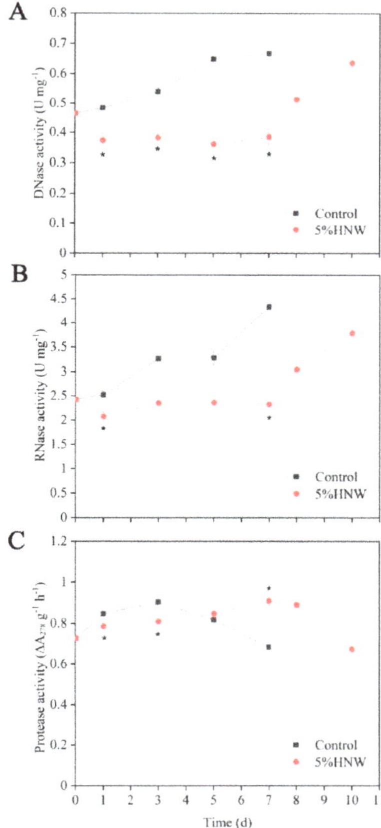

Figure 4. Effects of hydrogen nanobubble water (HNW) on the activities of DNase (**A**), RNase (**B**), and protease (**C**) in petals. Cut flowers were incubated in distilled water (control) and 5% HNW for 3 d (changed daily), and then, in distilled water, which was replaced daily until the end of experiments. Values were expressed as the means $\pm$ SE. Three samples per replicate were selected, and total samples in triplicate were 9 (3 $\times$ 3) for each treatment at each time point. Asterisks indicated significant differences at $p < 0.05$ according to t-test.

3. Discussion

It has been found that HRW can prolong the vase life of cut flowers, such as rose [21], lily [22], *Lisianthus* [23], and carnation [24]. Consistently, the experiments showed that 10% HRW also prolonged the vase life of cut carnation "Pink Diamond" flowers (Figure 2A). However, the residence time of H_2 in HRW was commonly shorter, with its half-life in water being about 100 min (Figure 1). Although MgH_2 as a soil H_2-storage material was suggested to be an alternative source for H_2 delivery due to the improvement of the solubility and residence time of dissolved H_2 [25], MgH_2 alone could prolong vase life by less than 30%, which was similar with HRW [26]. However, excess magnesium can cause symptoms resembling those of calcium deficiency and decrease the growth of rice and *Echinochloa* [32]. Similarly, in animals, excess magnesium is detrimental to the skeletal growth and development [33]. Therefore, there should be environmental and health concerns regarding magnesium when MgH_2 is widely used.

Since nanobubbles have unique properties with high internal pressure and negatively charged surface, these can improve the solubility and residence time of gases in liquid [28]. Nanobubbles have been used in water treatment [28] and soil remediation [34], and they also can promote the growth of animals, plants, and microbes [35–37]. Although several studies reported that HNW might optimize the composition of gut microbiota in mice [38] and decrease copper toxicity to *Daphnia magna* [30], the effect of HNW on postharvest preservation of cut flowers has not been reported.

As expected, in this study, HNW produced by a hydrogen nanobubble aerator also extended the residence time of H_2, and its half-life was about 150 min, in comparison with conventional HRW (about 100 min; Figure 1).

Among different doses of HNW, 5% HNW had the optimal effect on prolonging the vase life of cut carnation flowers, even better than 10% HRW (Figure 2A). Particularly, 5% HNW prolonged vase life by 51%, which was also larger than the effect of MgH_2 alone (prolonged vase life by 27% [25] and 29% [26]), even close to the combination of MgH_2 and citrate buffer solution (prolonged vase life by 52% [25]). Besides, HNW may be a good solvent for other additives to achieve a better preservation effect due to its simple composition.

Similar to the previous studies using higher content of HRW [21,23], 50% HNW cannot delay flower senescence, which may be attributable to its hypoxic effects [13,39], confirming that the effect of HNW on cut flower senescence was dose dependent in a specific range. Correlating with the changes in vase life, fresh weight, and flower diameter (Figure 2B–D), as well as the above H_2 content and residence time (Figure 1), it was suggested that 5% HNW significantly improved the availability of H_2 and prolonged the vase life of the flowers.

More importantly, the application of HNW is accordingly considered as a superior H_2 delivery method, and molecular hydrogen is completely harmless to environment and also beneficial to human health [40]. With the increasing home use of H_2 generator, the consumption cost of H_2 is reducing to as low as about 10 ¢/mg (https://h2hubb.com/2020/12/08/what-are-the-best-hydrogen-water-generators/, last accessed on 12 August 2021), about 10 ¢/1 L HNW.

Therefore, HNW may have wide-spread application, not only improving people's health, but also keeping cut flower fresh, thus making life beautiful.

Since water deficit results in flower wilting, maintaining cell turgor could delay cut flower senescence and improve their vase life [1]. It has been demonstrated that H_2 could enhance water conservation of alfalfa seedling leaves under drought stress [41] and rice root upon boron stress [42], as well as maintain a high level of RWC in cut lily and rose flowers during the vase period [21]. We previously observed that MgH_2- and HRW-supplied H_2 could delay RWC reduction in cut carnation flowers [26]. In the present study, HNW had a similar effect to RWC, resulting in alleviation of carnation petal wilting (Figures 2A and 3A).

ROS levels rise during flower senescence, which are highly detrimental for the protein stability and membrane integrity, thus contributing to the cell death and hastening flower senescence [43]. In this study, HNW reduced ROS accumulation induced by senescence, thus maintaining the membrane integrity (Figure 3B–D). These results were consistent with previous studies of the effects of ordinary HRW on cut rose, lily, and *Lisianthus* flowers [21,23]. The enhanced antioxidant capacity by H_2 was previously proposed as a primary mechanism in plant response against different stresses [12,13,27] and postharvest preservation of fruits [18] and flowers [21,23]. In a previous study, it was confirmed that at similar H_2 contents, HNW showed higher antioxidant activity than HRW without nanobubbles [29].

Both nucleic acid and protein degradation, resulted from the increased activities of nucleases and protease, play important roles in flower senescence [43]. In this study, the trends towards increased activities of DNase, RNase, and protease were observed during the vase life of cut carnation flowers (Figure 4). Interestingly, HNW decreased or delayed the activities of above three enzymes. Combined with data of cell death (Figure 3E) and phenotypes (Figure 2), we further propose that initial inhibition of the activities of nucleases and protease induced by HNW may partially contribute to alleviate cell death, thus delaying senescence and prolonging the vase life of cut flowers.

In conclusion, the present results clearly showed that, compared with HRW, the supply of HNW-mediated H_2 increased availability of H_2, which has a greater potential for application in horticulture. Furthermore, they also demonstrated remarkable roles of HNW in prolonging the vase life of cut flowers by reducing ROS accumulation and inhibiting the activities of nucleases and protease. These findings are expected to open a new window for low-carbon agriculture since H_2 has unique properties of renewable and zero greenhouse gas emissions on combustion.

4. Materials and Methods

4.1. Preparation of Hydrogen Nanobubble Water and Hydrogen-Rich Water

HNW was produced by a hydrogen nanobubble water generator (HIM-22; Guangdong Cawolo Health Technology Co., Ltd., Foshan, Guangdong, China). H_2 produced from water electrolysis was infused into 500 mL distilled water by a nanobubble aerator for 30 min. Conventional HRW was obtained by a H_2 generator (SHC-300; Saikesaisi Hydrogen Energy Co., Ltd., Jinan, Shandong, China), according to the previous method [23]. H_2 was bubbled into 500 mL distilled water at a rate of 150 mL min^{-1} for 30 min. The freshly prepared HNW/HRW (1 mg mL^{-1} and pH 8.6 $\pm$ 0.4/0.8 mg L^{-1} and 8.4 $\pm$ 0.3) was defined as 100% saturation HNW/HRW. Afterwards, 100% HNW or 100% HRW was immediately diluted to required concentration (1%, 5%, 10%, 50%, or 10% (*v/v*)), respectively. The concentrations of dissolved H_2 were measured by a portable dissolved hydrogen meter (CT-8023; Shenzhen Kedida Electronics Co., Ltd., Shenzhen, Guangdong, China; calibrated by gas chromatography). The mean diameter of H_2 nanobubbles in HNW was about 300 nm (determined by the NS300, Malvern Panalytical, Britain).

4.2. Plant Material and Treatments

Fresh cut carnation "Pink Diamond" flowers (within 1 d after harvest) were purchased from Hanzhongmen Flower Market (Nanjing, Jiangsu Province, China) and immediately transferred to the laboratory within 1 h. The cut carnations with the same degree of openness (the petals elongated vertically) and no mechanical damage were selected and placed in distilled water for 4 h. Afterwards, the stems were cut to a length of 25 cm under the water and the uppermost two leaves were kept.

Subsequently, cut carnations were incubated in distilled water (control), 1%, 5%, 10%, and 50% HNW and 10% HRW for 3 d (changed daily), and then, in distilled water, which was replaced daily until the end of experiments. Since 5% HNW showed the most obvious effect; it was used for the subsequent physiological and biochemical experiments. During the vase period, cut carnations were placed in an incubator at 25 °C and 80–85% relative humidity under a 12 h light/12 h dark photoperiod.

4.3. Measurement of Vase Life, Fresh Weight, and Flower Diameter

The vase life of cut carnation flower was calculated as the number of days from the day when flowers were placed in the vase solution until the day that 50% flowers wilted or had bent-neck (bent-neck angle greater than 45°). The fresh weight of cut flowers was measured daily, and relative fresh weight (RFW) was calculated as everyday fresh weight of cut flower against the initial day (0 d). Besides, flower diameter was determined as the maximum diameter of each flower and measured by using a caliper. There were three replicates and three flowers per each. Experiments were conducted in triplicate.

4.4. Determination of Water Content and Electrolyte Leakage of Petals

Relative water content (RWC) and electrolyte leakage of petals were measured according to the previous methods [21]. The fresh petals were weighed (W_f) and immersed in distilled water for 6 h at room temperature. Then, the turgid petals were dried and weighed (W_e). The turgid petals were oven dried at 80 °C to a constant weight and weighed again (W_d). RWC was calculated by the formula: RWC (%) = $[(W_f - W_d)/(W_e - W_d)] \times 100$.

Petals (0.2 g) were punched into 1 cm (diameter) discs and immersed in tubes with 20 mL distilled water for 4 h at room temperature after vacuuming for 30 min. Then, the initial conductivity (C_0) was determined. After incubating in boiling water for 15 min and cooling to room temperature, the conductivity (C) was determined again. The electrolyte leakage was calculated by the formula: electrolyte leakage (%) = $(C_0/C) \times 100$.

For determination of above two parameters, three flowers per replicate were selected, and total flowers in triplicate were 9 (3×3) for each treatment at each time point.

4.5. Histochemical Staining

ROS (H_2O_2 and O_2) accumulation was detected by DAB and NBT staining, respectively, according to the methods described previously [25,44]. The carnation petals were incubated in 0.1% (*w/v*) DAB or 0.1% (*w/v*) NBT solution for 12 h or 2 h in the dark at room temperature, respectively.

Trypan blue can only penetrate the membranes of dead cells, resulting in staining. The status of cell death in petals was detected by trypan blue staining according to previous method with minor modifications [45]. The petals were incubated in 2.5 mg mL^{-1} trypan blue solution for 1 h at room temperature. After washing extensively, the petals were photographed. Staining was performed with three flowers per replicate, and total flowers in triplicate were 9 (3×3) for each treatment at each time point.

4.6. Assays of Enzymatic Activity

Samples of 0.5 g fresh petals were homogenized in 3 mL of 0.1 M precooled acetic acid–sodium acetate buffer (pH 5.5) containing 1.1% polyvinylpyridoxone (PVP) in ice bath. The mixture was centrifuged at 12,000 rpm at 4 °C for 10 min. The supernatant was used for DNase and RNase activity assay according to the colorimetric method described previously [46]. For the determination of DNase activity, 200 μL of enzyme extract with 200 μL denatured calf thymus DNA (1 mg mL^{-1}) were incubated in 37 °C water bath for 1 h. Afterwards, the reaction was terminated by adding 95% (*v/v*) ethyl alcohol and settled at −20 °C for 12 h. The mixture was centrifuged at 12,000 rpm at 4 °C for 10 min. The absorbance of the supernatant was measured at 260 nm with a blank incubation without enzyme. One unit of enzyme activity was defined as the amount of enzyme causing an increased absorbance of 1.0 in 1 h at 260 nm. RNase activity was determined similarly with the following modification: yeast RNA (10 mg mL^{-1}) was used instead of calf thymus DNA (1 mg mL^{-1}). Enzyme activity was expressed as U mg^{-1} protein. Protein content was determined according to the Bradford method with BSA as a standard [47].

According to previous method [48], protease activity was determined by using hemoglobin as the substrate. The reaction mixture (0.2 mL 2% (*w/v*) denatured hemoglobin solution, 0.2 mL acetic acid–sodium acetate buffer (0.1 M, pH 5.2), and 0.2 mL crude extract) was incubated in 37 °C water bath for 1 h. The reaction was terminated by the addition of 0.8 mL 7.5% (*w/w*) trichloroacetic acid (TCA) and settled at 4 °C for 30 h. After centrifugation, the absorbance of the supernatant was measured at 278 nm against control that was added with TCA before reaction. Enzyme activity was expressed as ΔA_{278} g^{-1} (fresh weight) h^{-1}.

For activity assays of above three enzymes, three samples per replicate were selected, and total samples in triplicate were 9 (3×3) for each treatment at each time point.

4.7. Statistical Analysis

All values are expressed as the mean $\pm$ SE from three independent experiments for each treatment. Statistical analysis was performed by using SPSS 22.0 software (IBM Corporation, Armonk, NY, USA). The significant difference among treatments were analyzed by Duncan's multiple range test or *t*-test ($p < 0.05$ or 0.01).

Author Contributions: L.L. and W.S. conceived and designed the experiments. L.L. and Q.Y. performed the research. L.L. and Q.Y. analyzed the data. L.L. and Q.Y. wrote the original draft. L.L., T.Z., P.C., S.X., and W.S. reviewed and edited the paper. All authors have read and agreed to the published version of the manuscript.

Funding: This work was supported the Foshan Agriculture Science and Technology Project (Foshan City Budget, No. 140, 2019) and the Funding from Center of Hydrogen Science, Shanghai Jiao Tong University, China.

Institutional Review Board Statement: Not applicable.

Informed Consent Statement: Not applicable.

Data Availability Statement: All data, models, and code generated or used during the study appear in the submitted article.

Conflicts of Interest: The authors declare that they have no competing interests.

Abbreviations

Abbreviations
DAB	3,3-diaminobenzidine
RFW	relative fresh weight
H_2	hydrogen gas
HNW	hydrogen nanobubble water
HRW	hydrogen-rich water
H_2O_2	hydrogen peroxide
NBT	nitro blue tetrazolium
O_2^-	superoxide anions
PVP	polyvinylpyridoxone
RFW	relative fresh weight
ROS	reactive oxygen species
RWC	relative water content
SE	standard error

References

1. van Doorn, W.G.; Woltering, E.J. Physiology and molecular biology of petal senescence. *J. Exp. Bot.* **2008**, *59*, 453–480. [CrossRef] [PubMed]
2. Naing, A.H.; Lee, K.; Arun, M.; Lim, K.B.; Kim, C.K. Characterization of the role of sodium nitroprusside (SNP) involved in long vase life of different carnation cultivars. *BMC Plant Biol.* **2017**, *17*, 149. [CrossRef] [PubMed]
3. Rabiza-Świder, J.; Skutnik, E.; Jędrzejuk, A.; Rochala-Wojciechowska, J. Nanosilver and sucrose delay the senescence of cut snapdragon flowers. *Postharvest Biol. Technol.* **2020**, *165*, 111165. [CrossRef]

4. Langston, B.J.; Bai, S.; Jones, M.L. Increases in DNA fragmentation and induction of a senescence-specific nuclease are delayed during corolla senescence in ethylene-insensitive (etr1-1) transgenic petunias. *J. Exp. Bot.* **2005**, *56*, 15–33. [CrossRef]

5. Panavas, T.; LeVangie, R.; Mistler, J.; Reid, P.D.; Rubinstein, B. Activities of nucleases in senescing daylily petals. *Plant Physiol. Biochem.* **2000**, *38*, 837–843. [CrossRef]

6. Eason, J.R.; Ryan, D.J.; Pinkney, T.T.; O'Donoghue, E.M. Programmed cell death during flower senescence: Isolation and characterization of cysteine proteinases from *Sandersonia aurantiaca*. *Funct. Plant Biol.* **2002**, *29*, 1055–1064. [CrossRef] [PubMed]

7. Pak, C.; van Doorn, W.G. Delay of *Iris* flower senescence by protease inhibitors. *New Phytol.* **2005**, *165*, 473–480. [CrossRef]

8. Ohsawa, I.; Ishikawa, M.; Takahashi, K.; Watanabe, M.; Nishimaki, K.; Yamagata, K.; Katsura, K.; Katayama, Y.; Asoh, S.; Ohta, S. Hydrogen acts as a therapeutic antioxidant by selectively reducing cytotoxic oxygen radicals. *Nat. Med.* **2007**, *13*, 688–694. [CrossRef]

9. Ohta, S. Recent progress toward hydrogen medicine: Potential of molecular hydrogen for preventive and therapeutic applications. *Curr. Pharm. Des.* **2011**, *17*, 2241. [CrossRef]

10. Renwick, G.M.; Giumarro, C.; Siegel, S.M. Hydrogen metabolism in higher plants. *Plant Physiol.* **1964**, *39*, 303–306. [CrossRef] [PubMed]

11. Zeng, J.; Zhang, M.; Sun, X. Molecular hydrogen is involved in phytohormone signaling and stress responses in plants. *PLoS ONE* **2013**, *8*, e71038. [CrossRef]

12. Xie, Y.; Mao, Y.; Lai, D.; Zhang, W.; Shen, W. H_2 enhances Arabidopsis salt tolerance by manipulating ZAT10/12-mediated antioxidant defence and controlling sodium exclusion. *PLoS ONE* **2012**, *7*, e49800. [CrossRef]

13. Xie, Y.; Mao, Y.; Zhang, W.; Lai, D.; Wang, Q.; Shen, W. Reactive oxygen species-dependent nitric oxide production contributes to hydrogen-promoted stomatal closure in Arabidopsis. *Plant Physiol.* **2014**, *165*, 759–773. [CrossRef]

14. Su, J.; Yang, X.; Shao, Y.; Chen, Z.; Shen, W. Molecular hydrogen–induced salinity tolerance requires melatonin signalling in *Arabidopsis thaliana*. *Plant Cell Environ.* **2021**, *44*, 476–490. [CrossRef]

15. Cui, W.; Yao, P.; Pan, J.; Dai, C.; Cao, H.; Chen, Z.; Zhang, S.; Xu, S.; Shen, W. Transcriptome analysis reveals insight into molecular hydrogen-induced cadmium tolerance in alfalfa: The prominent role of sulfur and (homo)glutathione metabolism. *BMC Plant Biol.* **2020**, *20*, 58. [CrossRef]

16. Cao, Z.; Duan, X.; Yao, P.; Cui, W.; Cheng, D.; Zhang, J.; Jin, Q.; Chen, J.; Dai, T.; Shen, W. Hydrogen gas is involved in auxin-induced lateral root formation by modulating nitric oxide synthesis. *Int. J. Mol. Sci.* **2017**, *18*, 2084. [CrossRef] [PubMed]

17. Zhang, X.; Su, N.; Jia, L.; Tian, J.; Li, H.; Huang, L.; Shen, Z.; Cui, J. Transcriptome analysis of radish sprouts hypocotyls reveals the regulatory role of hydrogen-rich water in anthocyanin biosynthesis under UV-A. *BMC Plant Biol.* **2018**, *18*, 227. [CrossRef] [PubMed]

18. Hu, H.; Li, P.; Song, Y.; Gu, R. Hydrogen-rich water delays postharvest ripening and senescence of kiwifruit. *Food Chem.* **2014**, *156*, 100–109. [CrossRef] [PubMed]

19. Hu, H.; Zhao, S.; Li, P.; Shen, W. Hydrogen gas prolongs the shelf life of kiwifruit by decreasing ethylene biosynthesis. *Postharvest Biol. Technol.* **2018**, *135*, 123–130. [CrossRef]

20. Zhang, Y.; Zhao, G.; Cheng, P.; Yan, X.; Li, Y.; Cheng, D.; Wang, R.; Chen, J.; Shen, W. Nitrite accumulation during storage of tomato fruit as prevented by hydrogen gas. *Int. J. Food Prop.* **2019**, *22*, 1425–1438. [CrossRef]

21. Ren, P.; Jin, X.; Liao, W.; Wang, M.; Niu, L.; Li, X.; Xu, X.; Zhu, Y. Effect of hydrogen-rich water on vase life and quality in cut lily and rose flowers. *Hortic. Environ. Biotechnol.* **2017**, *58*, 576–584. [CrossRef]

22. Huo, J.; Huang, D.; Zhang, J.; Fang, H.; Wang, B.; Wang, C.; Ma, Z.; Liao, W. Comparative proteomic analysis during the involvement of nitric oxide in hydrogen gas-improved postharvest freshness in cut lilies. *Int. J. Mol. Sci.* **2018**, *19*, 3955. [CrossRef] [PubMed]

23. Su, J.; Nie, Y.; Zhao, G.; Cheng, D.; Wang, R.; Chen, J.; Zhang, S.; Shen, W. Endogenous hydrogen gas delays petal senescence and extends the vase life of lisianthus cut flowers. *Postharvest Biol. Technol.* **2019**, *147*, 148–155. [CrossRef]

24. Cai, M.; Du, H. Effects of hydrogen-rich water pretreatment on vase life of carnation (*Dianthus caryophyllus*) cut flowers. *J. Shanghai Jiaotong Univ.* **2015**, *33*, 41–45.

25. Li, L.; Liu, Y.; Wang, S.; Zou, J.; Ding, W.; Shen, W. Magnesium hydride-mediated sustainable hydrogen supply prolongs the vase life of cut carnation flowers via hydrogen sulfide. *Front. Plant Sci.* **2020**, *11*, 595376. [CrossRef]

26. Li, Y.; Li, L.; Wang, S.; Liu, Y.; Zou, J.; Ding, W.; Du, H.; Shen, W. Magnesium hydride acts as a convenient hydrogen supply to prolong the vase life of cut roses by modulating nitric oxide synthesis. *Postharvest Biol. Technol.* **2021**, *177*, 111526. [CrossRef]

27. Li, L.; Lou, W.; Kong, L.; Shen, W. Hydrogen commonly applicable from medicine to agriculture: From molecular mechanisms to the field. *Curr. Pharm. Des.* **2021**, *27*, 747–759. [CrossRef]

28. Temesgen, T.; Bui, T.T.; Han, M.; Kim, T.; Park, H. Micro and nanobubble technologies as a new horizon for water-treatment techniques: A review. *Adv. Colloid Interface Sci.* **2017**, *246*, 40–51. [CrossRef] [PubMed]

29. Liu, S.; Oshita, S.; Thuyet, D.Q.; Saito, M.; Yolimoto, T. Antioxidant activity of hydrogen nanobubbles in water with different reactive oxygen species both in vivo and in vitro. *Langmuir* **2018**, *34*, 11878–11885. [CrossRef]

30. Fan, W.; Zhang, Y.; Liu, S.; Li, X.; Li, J. Alleviation of copper toxicity in *Daphnia magna* by hydrogen nanobubble water. *J. Hazard. Mater.* **2020**, *389*, 122155. [CrossRef]

31. Arora, A.; Singh, V.P. Polyols regulate the flower senescence by delaying programmed cell death in *Gladiolus*. *J. Plant Biochem. Biotechnol.* **2006**, *15*, 139–142. [CrossRef]

32. Kobayashi, H.; Masaoka, Y.; Sato, S. Effects of excess magnesium on the growth and mineral content of rice and *Echinochloa*. *Crop. Physiol. Ecol.* **2005**, *8*, 38–43. [CrossRef]
33. Bao, S.F.; Zhao, L.; Li, Z.; Cong, T. The influence of different dietary magnesium levels on the metabolism of calcium, phosphorus and magnesium in growing rats. *Trace Elem. Electrolytes* **2000**, *17*, 92–96.
34. Kim, D.; Han, J. Remediation of copper contaminated soils using water containing hydrogen nanobubbles. *Appl. Sci.* **2020**, *10*, 2185. [CrossRef]
35. Ebina, K.; Shi, K.; Hirao, M.; Hashimoto, J.; Kawato, Y.; Kaneshiro, S.; Morimoto, T.; Koizumi, K.; Yoshikawa, H. Oxygen and air nanobubble water solution promote the growth of plants, fishes, and mice. *PLoS ONE* **2013**, *8*, e65339. [CrossRef]
36. Ahmed, A.K.A.; Shi, X.; Hua, L.; Manzueta, L.; Qing, W.; Marhaba, T.; Zhang, W. Influences of air, oxygen, nitrogen, and carbon dioxide nanobubbles on seed germination and plant growth. *J. Agric. Food Chem.* **2018**, *66*, 5117–5124. [CrossRef] [PubMed]
37. Zhu, J.; Wakisaka, M. Effect of air nanobubble water on the growth and metabolism of *Haematococcus lacustris* and *Botryococcus braunii*. *J. Nutr. Sci. Vitaminol.* **2019**, *65*, S212–S216. [CrossRef]
38. Guo, Z.; Hu, B.; Han, H.; Lei, Z.; Shimizu, K.; Zhang, L.; Zhang, Z. Metagenomic insights into the effects of nanobubble water on the composition of gut microbiota in mice. *Food Funct.* **2020**, *11*, 7175–7182. [CrossRef]
39. Su, J.; Zhang, Y.; Nie, Y.; Cheng, D.; Wang, R.; Hu, H.; Chen, J.; Zhang, J.; Du, Y.; Shen, W. Hydrogen-induced osmotic tolerance is associated with nitric oxide-mediated proline accumulation and reestablishment of redox balance in alfalfa seedlings. *Environ. Exp. Bot.* **2018**, *147*, 249–260. [CrossRef]
40. Matei, N.; Camara, R.; Zhang, J.H. Emerging mechanisms and novel applications of hydrogen gas therapy. *Med. Gas. Res.* **2018**, *8*, 98–102.
41. Chen, Y.; Wang, M.; Hu, L.; Liao, W.; Dawuda, M.M.; Li, C. Carbon monoxide is involved in hydrogen gas-induced adventitious root development in cucumber under simulated drought stress. *Front. Plant Sci.* **2017**, *8*, 128. [CrossRef] [PubMed]
42. Wang, Y.; Duan, X.; Xu, S.; Wang, R.; Ouyang, Z.; Shen, W. Linking hydrogen-mediated boron toxicity tolerance with improvement of root elongation, water status and reactive oxygen species balance: A case study for rice. *Ann. Bot.* **2016**, *118*, 1279–1291. [CrossRef] [PubMed]
43. Ahmad, S.S.; Tahir, I. How and why of flower senescence: Understanding from models to ornamentals. *Indian J. Plant Physiol.* **2016**, *21*, 446–456. [CrossRef]
44. Su, N.; Wu, Q.; Liu, Y.; Cai, J.; Shen, W.; Xia, K.; Cui, J. Hydrogen-rich water reestablishes ROS homeostasis but exerts differential effects on anthocyanin synthesis in two varieties of radish sprouts under UV-A irradiation. *J. Agric. Food Chem.* **2014**, *62*, 6454–6462. [CrossRef]
45. Bowling, S.A.; Clarke, J.D.; Liu, Y.; Klessig, D.F.; Dong, X. The cpr5 mutant of Arabidopsis expresses both NPRl-dependent and NPR1-independent resistance. *Pant Cell* **1997**, *9*, 1573–1584.
46. Yang, L.; Guo, A.; Wang, P. Research on the nucleic acid contents and nuclease activated for leaves of wheat albescent line during the albescent period. *Acta Agric. Boreali Occident. Sin.* **2001**, *10*, 32–35.
47. Bradford, M.M. A rapid and sensitive method for the quantitation of microgram quantities of protein utilizing the principle of protein-dye binding. *Anal. Biochem.* **1976**, *72*, 248–254. [CrossRef]
48. Zhang, Z.; Rui, Q.; Xu, L. Relationship between endopeptidase and H_2O_2 during the aging of wheat leaf. *Acta Bot. Sin.* **2001**, *42*, 127–131.

Article

Molecular Hydrogen Increases Quantitative and Qualitative Traits of Rice Grain in Field Trials

Pengfei Cheng [1], Jun Wang [1], Zhushan Zhao [1], Lingshuai Kong [1], Wang Lou [1], Tong Zhang [1], Dedao Jing [2], Julong Yu [2], Zhaolin Shu [2], Liqin Huang [3], Wenjiao Zhu [1], Qing Yang [1] and Wenbiao Shen [1,4,*]

[1] Laboratory Center of Life Sciences, College of Life Sciences, Nanjing Agricultural University, Nanjing 210095, China; 2020216037@njau.edu.cn (P.C.); 2020116099@stu.njau.edu.cn (J.W.); 2021816131@stu.njau.edu.cn (Z.Z.); 2018116100@njau.edu.cn (L.K.); 2018116099@njau.edu.cn (W.L.); 2020116098@stu.njau.edu.cn (T.Z.); zhuwenjiao@njau.edu.cn (W.Z.); qyang19@njau.edu.cn (Q.Y.)
[2] Zhenjiang Institute of Agricultural Science of the Ning-Zhen Hilly District, Jurong 212400, China; jingdedao@163.com (D.J.); yujulong@126.com (J.Y.); shuzl2005@163.com (Z.S.)
[3] College of Sciences, Nanjing Agricultural University, Nanjing 210095, China; lqhuangs@njau.edu.cn
[4] Center of Hydrogen Science, Shanghai Jiao Tong University, Shanghai 200240, China
* Correspondence: wbshenh@njau.edu.cn

Abstract: How to use environmentally friendly technology to enhance rice field and grain quality is a challenge for the scientific community. Here, we showed that the application of molecular hydrogen in the form of hydrogen nanobubble water could increase the length, width, and thickness of brown/rough rice and white rice, as well as 1000-grain weight, compared to the irrigation with ditch water. The above results were well matched with the transcriptional profiles of representative genes related to high yield, including up-regulation of *heterotrimeric G protein β-subunit gene (RGB1)* for cellular proliferation, *Grain size 5 (GS5)* for grain width, *Small grain 1 (SMG1)* for grain length and width, *Grain weight 8 (GW8)* for grain width and weight, and down-regulation of negatively correlated gene *Grain size 3 (GS3)* for grain length. Meanwhile, although total starch content in white rice is not altered by HNW, the content of amylose was decreased by 31.6%, which was parallel to the changes in the transcripts of the amylose metabolism genes. In particular, cadmium accumulation in white rice was significantly reduced, reaching 52% of the control group. This phenomenon was correlated well with the differential expression of transporter genes responsible for Cd entering plants, including down-regulated *Natural resistance-associated macrophage protein (Nramp5)*, *Heavy metal transporting ATPase (HMA2* and *HMA3)*, and *Iron-regulated transporters (IRT1)*, and for decreasing Cd accumulation in grain, including down-regulated *Low cadmium (LCD)*. This study clearly showed that the application of molecular hydrogen might be used as an effective approach to increase field and grain quality of rice.

Keywords: amylose; cadmium; field quality; hydrogen-based agriculture; hydrogen nanobubble water; rice

Citation: Cheng, P.; Wang, J.; Zhao, Z.; Kong, L.; Lou, W.; Zhang, T.; Jing, D.; Yu, J.; Shu, Z.; Huang, L.; et al. Molecular Hydrogen Increases Quantitative and Qualitative Traits of Rice Grain in Field Trials. *Plants* **2021**, *10*, 2331. https://doi.org/10.3390/plants10112331

Academic Editor: John Hancock

Received: 10 October 2021
Accepted: 26 October 2021
Published: 28 October 2021

1. Introduction

Rice is typically milled from brown rice to white rice, the most-often consumed form of rice, and more than 3 billion people use rice as their main food, particularly in Asian, South-American, and African countries [1]. Since it is rich in proteins, carbohydrates, vitamins, biologically active compounds, and organic acids, rice is generally good for human health [2]. However, rice normally appears to have more absorption of cadmium (Cd), a very toxic heavy metal caused by soil contamination and acidification, compared to other major cereal crops. This could result in the accumulation of Cd in rice grains exceeding the maximum permissible limit [3]. Meanwhile, amylose content is a key determinant of eating quality of rice [4]. Therefore, avoiding excessive Cd accumulation in white rice and improving field and grain quality, especially breeding or producing low-amylose content rice, are not only important for consideration during rice production, but also a challenge for scientific community.

The improvement of rice yield and quality normally utilizes molecular genetic selection combined with hybridization, which is mainly dependent on various rice germplasm resources [5]. The improved germplasm is normally integrated with the application of chemical fertilizers and pesticides [6]. During the last 25 years, the usage of recombinant genetic methods has been academically confirmed to be more efficient and reliable [7–9], but this technology requires relatively difficult approval protocols and has faced reluctance from consumers. The use of excessive pesticides and fertilizers in fields is another problem, since these can easily cause serious environmental pollution [6].

Molecular hydrogen is generally applied in clean energy. Although hydrogen gas (H_2) is normally considered as a biologically inert gas, previous studies discovered that this gas could act as a therapeutic antioxidant in medicine [10]. Scientists have gradually realized that molecular hydrogen has anti-inflammatory, antioxidant, and anti-apoptotic effects in animal models and basic clinical research [11]. During the recent decade, the physiological functions of H_2 in plants have been discovered, including enhancing plant tolerance against abiotic stress [12–14], and promoting plant growth and development [15]. For above cases, H_2 might be integrated with other downstream gasotransmitters, including nitric oxide and hydrogen sulfide, as well as regulating some phytohormones [11,16]. In terms of postharvest storage period, previous results also revealed that H_2 could prolong the shelf life of some vegetables, fruits, and flowers, including kiwifruit [17] and daylily bud [18], as well as lisianthus [19], carnation [20], and Chinese chive [21] when hydrogen-rich water, H_2 fumigation, or magnesium hydride (a H_2-releasing material) were separately applied. Since there were numerous functions of H_2 observed in the vegetative growth stage and postharvest period, it is reasonable to deduce that molecular hydrogen might have significant influence in reproductive growth and seed developmental stages, both of which are very important for crop production, especially for rice.

Hydrogen-rich water (HRW) or saline is a typical form of molecular hydrogen used in plants and medicine, due to its feasibility and safety in laboratory [14]. While the low solubility and short residence time of H_2 limit its wide application in practice, especially in paddy field, previous results have revealed that nano-bubble technology could be applied in aquaculture [22] and environment [23], since this approach could increase the content of targeted gas in water, prolong the time of the gas remaining in liquid, and improve the utilization efficiency of targeted gas. A recent study further discovered that the hydrogen nanobubble water (HNW) could alleviate copper toxicity in *Daphnia magna* since it could increase the solubility and the residence time of H_2 in water, therefore efficiently enhancing antioxidant capacity [24]. For above reasons, HNW was used in our field experimental condition.

The objective of this work was to investigate whether and how HNW could influence field and grain quality traits of rice, compared to the ditch water treated control group. Related results further supported the idea that the field trait, the qualitative characters, and quality of polished white rice were ubiquitously improved by HNW. Importantly, amylose content and Cd accumulation in white rice were significantly reduced. The related molecular mechanisms were preliminarily evaluated in terms of the transcript profiles of molecular marker genes.

2. Results

2.1. Seed Morphology of Rice Crops Irrigated with HNW

The seed size of crops is an important trait of yield. The size of the grain influences many aspects of plant growth and development. In the trail experiment, we observed that the difference of length and width (in particularly) of rice grains (Figure 1A–D) harvested from plots either irrigated with HNW or ditch water (control) was clearly distinct. For example, the average length and width of the grains in HNW-treated group was about 11.4% and 15.1% greater than those from the control sample. Compared to grains harvested from the control, grain length-width ratio after irrigation with HNW was decreased by 4.3% (Figure 1E). Particularly, grain thickness showed pronounced improvement by 37.5% after HNW treatment (Figure 1F). The obvious increase in seed setting ratio was observed as well (Figure 1G). These results could be used to explain the significant increase in thousand-grain weight (about 23.8%; Figure 1H).

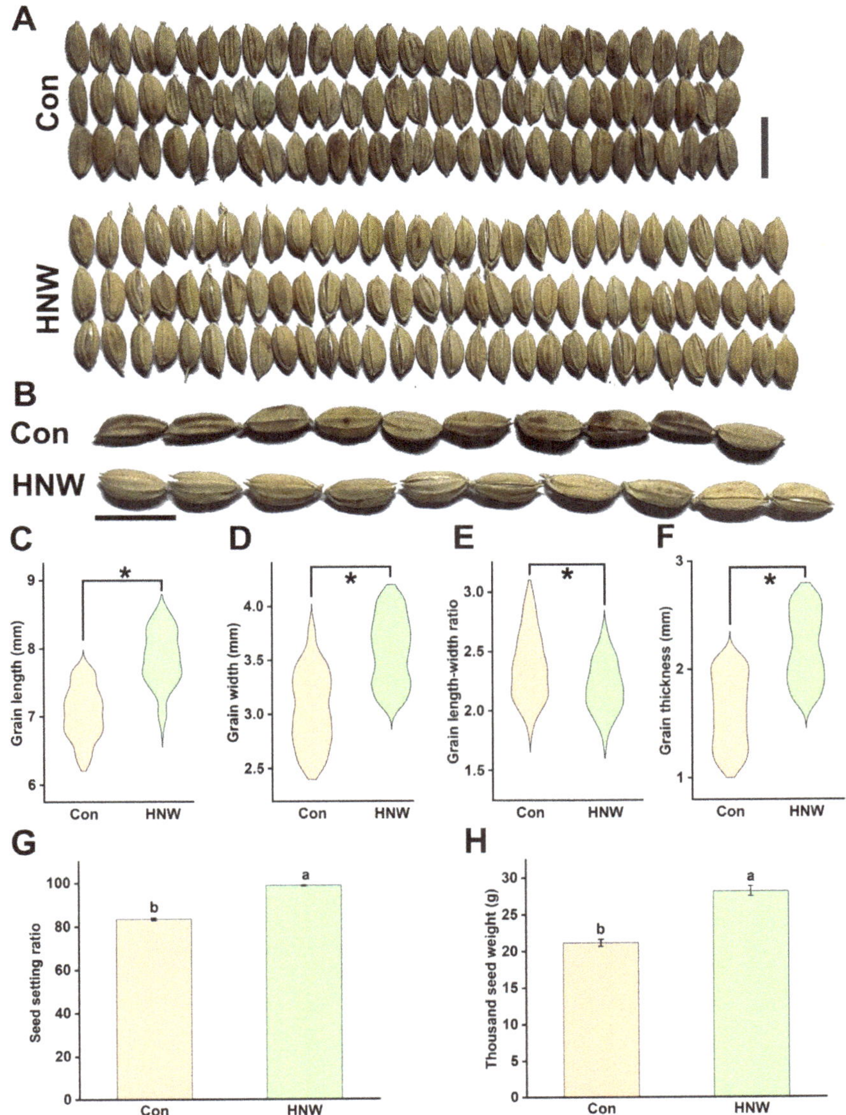

Figure 1. HNW positively influences the size and weight of grains. The photographs of representative 30 grains (**A**) and 10 grains (**B**) were taken (bar = 1 cm). Parameters of rice size, including grain length (**C**), width (**D**), length-width ratio (**E**), thickness (**F**), seed setting ratio (**G**), and thousand seed weight (**H**), were also provided. Asterisk indicates a significant difference between Con and HNW (n ≥ 1000, $p < 0.001$, two-way Student's *t*-test). Data are mean ± SD (n = 3). Bars with different letters were significantly different in comparison with Con at $p < 0.05$.

Molecular evidence revealed that the above changes achieved by HNW were well matched with the transcriptional profiles of representative molecular markers responsible for high yield in rice young panicles (Figure 2). These included up-regulated *heterotrimeric G protein β-subunit gene (RGB1)* for cellular proliferation [25], *Small grain 1 (SMG1)* for grain length and width [26], *Grain size 5 (GS5)* for grain width [27,28], and *Grain weight 8 (GW8)* for grain weight [29], after being irrigated with HNW. Meanwhile, the down-regulation of *Grain size 3 (GS3)*, negatively correlated with grain length [29], was observed after irrigation with HNW. Further results illustrated that the transcripts of representative genes related to the absorption of nitrogen (N), phosphorus (P), and potassium (K) in rice plants [30], including controlling N assimilation and transport, as well as P and K absorption, especially *Nitrate transporters 2.3 (NRT2.3)*, *Nitrite reductase (NiR)*, *ABC1 repressor 1 (ARE1)*, *Nin-like protein 4 (NLP4)*, and *Potassium transporter 1 (AKT1)*, were obviously increased in HNW-irrigated rice roots (Figure 3).

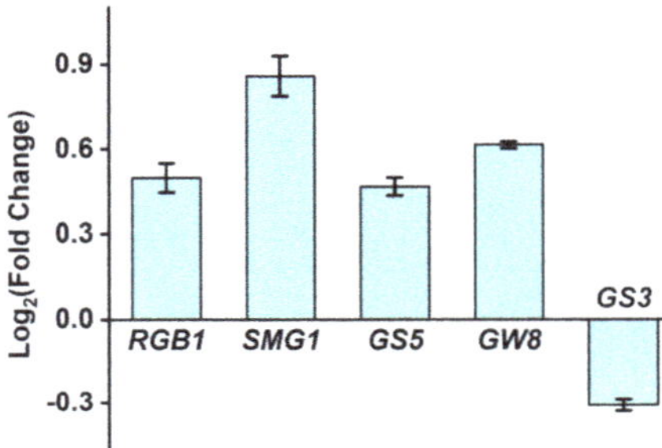

Figure 2. The results of gene expression in (young panicles) at the filling stage (5 September 2020) show that HNW may increase rice seed size. Then, transcripts of *RGB1*, *SMG1*, *GS5*, *GW8*, and *GS3* were analyzed by qPCR. Data are mean ± SD (n = 3).

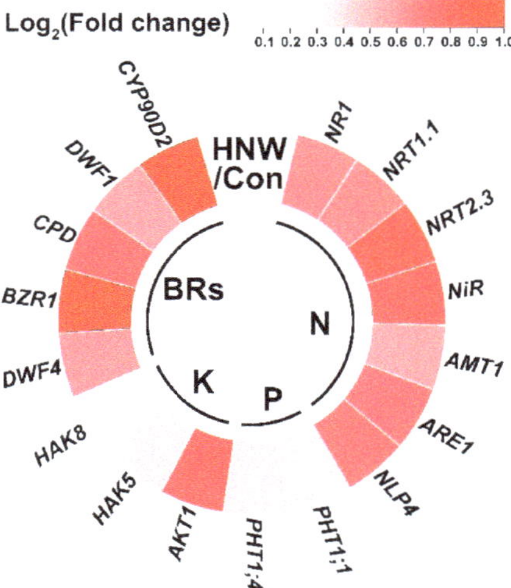

Figure 3. Molecular evidence showing that the results of rice seed size might be causally related to absorption of N, P, and K, as well as BRs signaling. Transcripts of genes related to absorption of N, P, and K in root tissues, and BRs signaling in young panicles were analyzed by qPCR. The samples were obtained on 5 September 2020. The different color means different log$_2$(fold change) of HNW contrast Con.

Since some of these genes, including *SMG1*, *GS5*, *NR1*, and *AMT1*, could be modulated by brassinolide (BRs) [26,27,31–33], so expression of specific genes related to BRs metabolism and signaling was further analyzed. As expected, the similar inducting profiles in *DWARF4* (*DWF4*), *Brassinosteroid signaling positive regulator* (*BZR1*), *Constitutive photomorphogenic dwarf* (*CPD*), *DWARF1* (*DWF1*), and *Cytochrome P450 90D2* (*CYP90D2*) mRNAs were observed in the HNW-treated rice young panicles (Figure 3). These changes might be used to explain the improvement of field and grain quantity in the HNW-irrigated group.

2.2. The Morphology of White Rice

After processes, the rice grains are polished to edible white rice. Similar to the changes in grains (Figure 1), the length and width (especially) of white rice were obviously increased after HNW irrigation, compared to those of the ditch water-irrigated control group (Figure 4A–C). Although the changes in length-width ratio of white rice was negatively influenced by HNW (Figure 4D), both the thickness (Figure 4E) and thousand-seed weight (Figure 4F) displayed distinct improvements in comparison with the control group. Together, we found that molecular hydrogen could increase quantitative and qualitative traits of rice grain in field trials.

2.3. Qualitative Characters of White Rice Irrigated with HNW

Ample evidence found that the qualitative characters of white rice were closely related to gel consistency, chalkiness rice rate, contents of protein, total starch, and amylose contents [4]. Subsequent results discovered that compare to the control group (83.1 ± 2.4 mm), HNW could increase the gel consistency to 91.4 ± 3.4 mm (Figure 5A), a relatively perfect support for higher quality of eaten rice [34]. Strikingly, the chalkiness rate of rice was obviously decreased by HNW (Figure 5B).

Subsequent results revealed that HNW irrigation could significantly reduce the total protein level (decreased by 19.8%; Figure 5C) and amylose content (decreased by 31.6%; Figure 5D) without altering the total starch content (Figure 5E). These are interesting findings. To probe the mechanism, some related genes controlling low amylose content in white rice, including *granule-bound starch synthase1* (*GBSS1*), *starch isomerase1* (*ISA1*), and *starch branching enzyme1/2* (*SBE1/2*), were analyzed (Figure 6). As anticipated, the changes in transcriptional expression of target genes related to amylose were well matched with the reduction in amylose content. These results clearly illustrated that HNW could decrease amylose synthesis.

Although our results discovered that HNW could reduce total protein content in white rice (Figure 5C), the content of glutelin, an important and major storage protein in the endosperm of rice, was not significantly altered in response to HNW, with respect to the control sample (Figure 7A). Contents of other grain storage proteins, including prolamin (especially), globulin, and albumin (Figure 7B–D), were obviously impaired by HNW irrigation. Moreover, no significant differences were observed in the contents of the vitamin B1 and B5 in white rice after being irrigated with HNW or in the control group (data are not shown).

2.4. HNW Influences Contents of the Metal Ions in White Rice

Further experiment was carried out to assess the possible link between element content of white rice and HNW irrigation. In our experimental conditions, the contents of Cd and tin (Sn) were reduced significantly by HNW, compared to those in control group (Figure 8A,B). The reduction in Cd accumulation by about 52% in white rice was very interesting since no significant difference in Cd content was discovered in soils sampled from the control and HNW-irrigated fields (data are not shown). Comparatively, we also noticed that HNW could increase the contents of some nutrient elements that are beneficial for both plants and humans, including phosphorus (P; Figure 8C), potassium (K; Figure 8D), magnesium (Mg; Figure 8E), and iron (Fe; Figure 8F) in white rice.

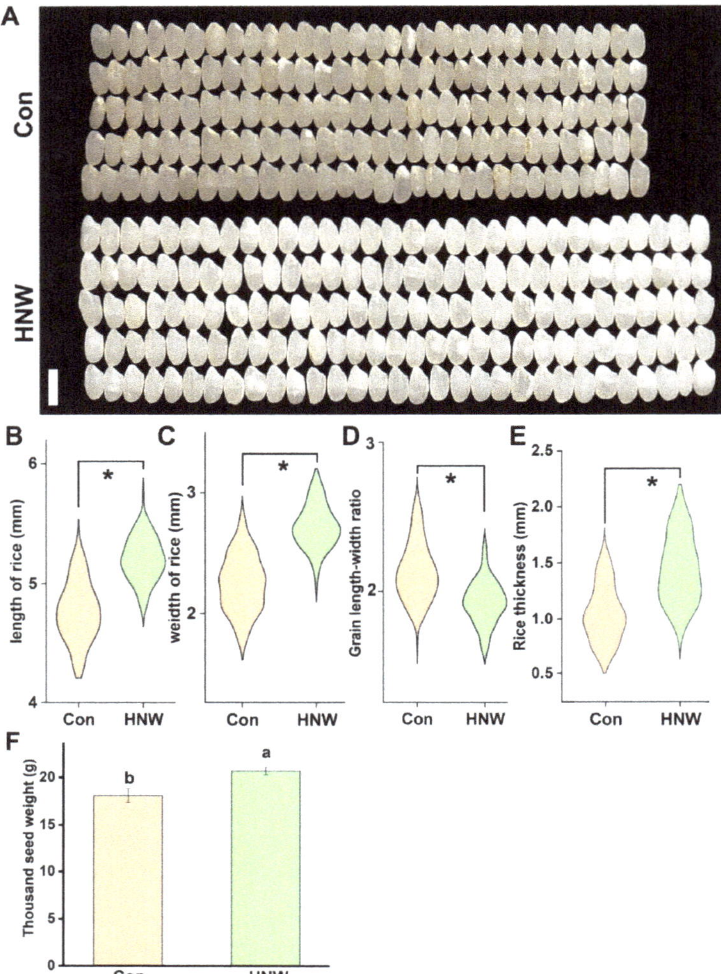

Figure 4. HNW positively influences the size and weight of white rice. The photographs of 30 white rice (**A**) were taken (bar = 0.5 cm). Parameters of white rice size, including length (**B**), width (**C**), length-width ratio (**D**), thickness (**E**), and thousand grain weight (**F**), were analyzed. Asterisk indicates a significant difference between Con and HNW (n $\geq$ 1000, $p < 0.001$, two-way Student's *t*-test). Data are mean $\pm$ SD (n = 3). Bars with different letters were significantly different in comparison with Con at $p < 0.05$.

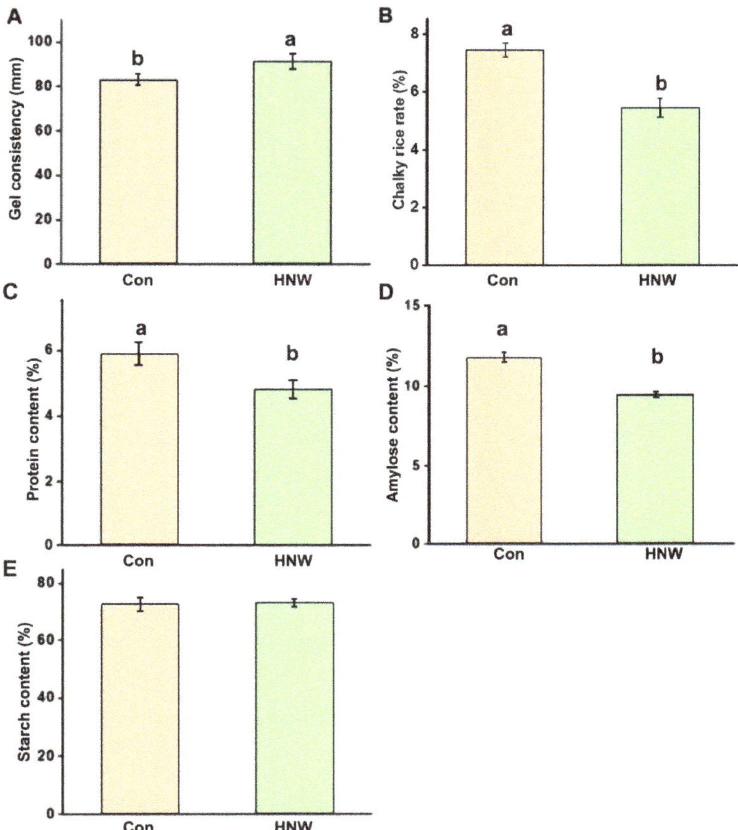

Figure 5. HNW positively influences qualitative characters of white rice. The gel consistency (**A**), chalky rice rate (**B**), contents of total protein (**C**), amylose (**D**), and starch (**E**) were analyzed. Data are mean ± SD (n = 3). Bars with different letters were significantly different in comparison with Con at $p < 0.05$.

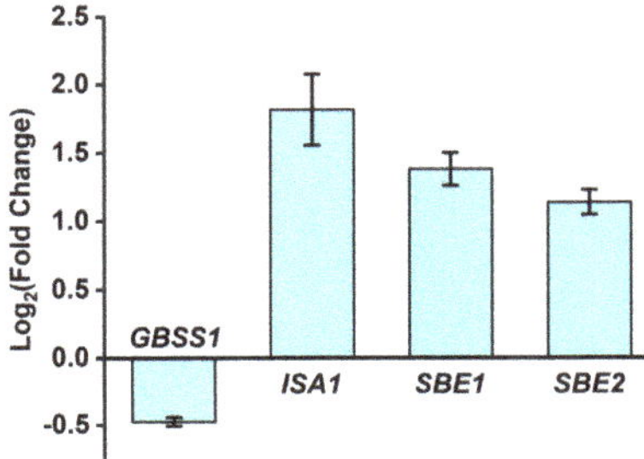

Figure 6. Changes of gene expression in leaves at the filling stage (5 September 2020) showing that HNW might decrease amylose accumulation. Transcripts of *GBSS1*, *ISA1*, *SBE1*, and *SBE2* were analyzed by qPCR. Data are mean ± SD (n = 3).

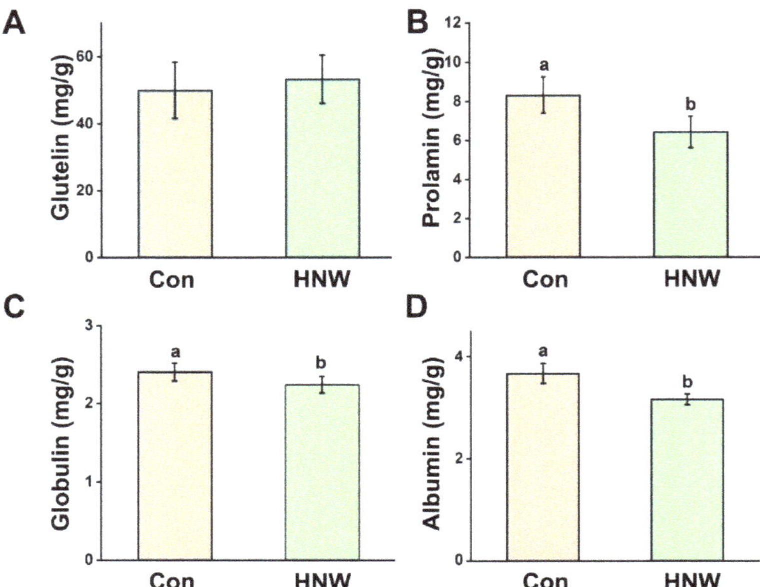

Figure 7. Composition of the crude protein in white rice. The contents of glutelin (**A**), prolamin (**B**), globulin (**C**), and albumin (**D**) were measured, respectively. Data are mean ± SD (n = 3). Bars with different letters were significantly different in comparison with Con at $p < 0.05$.

To gain insight into the molecular mechanism, the transcriptional abundance of genes responsible for reducing Cd content was analyzed in root tissues during the filling stage (Figure 9), and the change in gene expression could at least partly explain the variation of Cd content. For instance, *Natural resistance-associated macrophage protein* (*Nramp5*) [35], *Heavy metal transporting ATPase* (*HMA2 and HMA3*) [3,35], *Iron-regulated transporters* (*IRT1*) [3], and *Low cadmium* (*LCD*) [36], all of which were responsible for governing the entry and accumulation of Cd in plants, were remarkably down-regulated by HNW irrigation. Consistent with the changes in Cd content of white rice with or without HNW irrigation (Figure 8A), these molecular evidences clearly supported the idea that HNW irrigation was closely associated with the reduction of Cd accumulation via modulating transcriptional regulation.

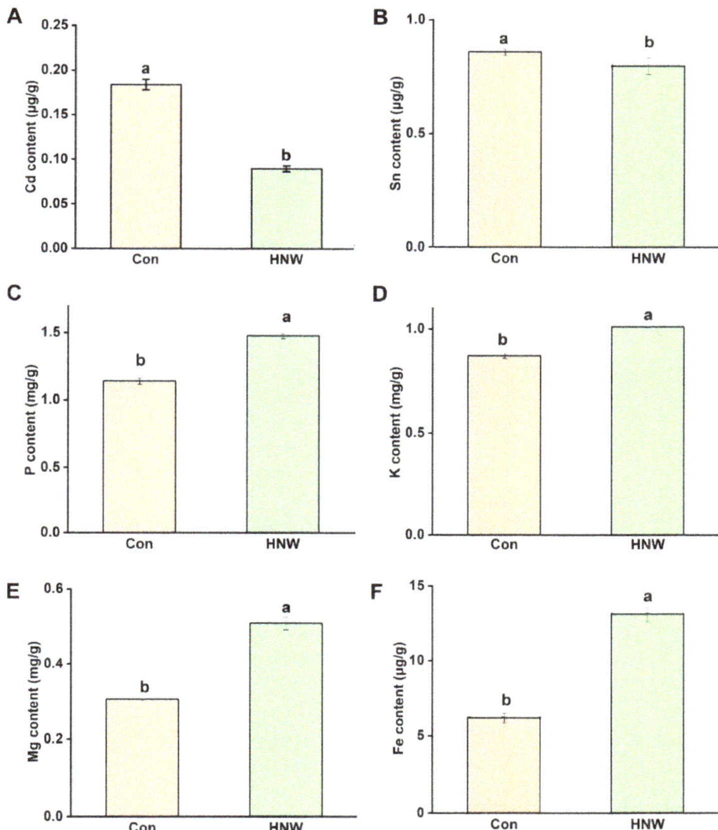

Figure 8. HNW affects the absorption of heavy metal and some nutrient element in white rice. The heavy metal content was analyzed, including Cd (**A**) and Sn (**B**). Meanwhile some nutrient elements, including phosphorus (**C**), potassium (**D**), magnesium (**E**), and iron (**F**), in white rice were further determined. Data are mean $\pm$ SD (n = 3). Bars with different letters were significantly different in comparison with Con at $p < 0.05$.

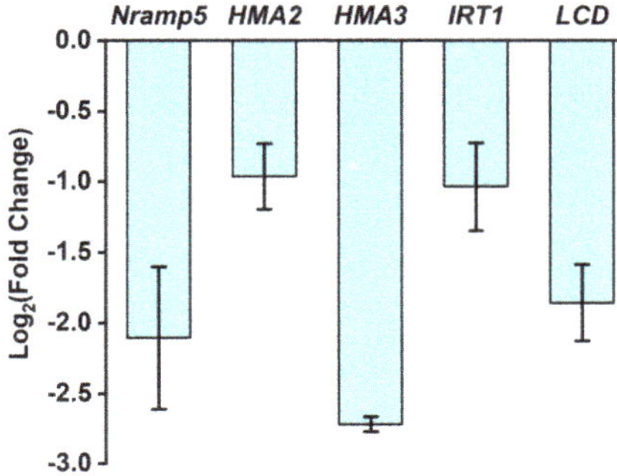

Figure 9. The results of gene expression in roots at the filling stage (5 September 2020) showing that HNW could decrease cadmium accumulation. Transcripts of *Nramp5*, *HMA2*, *HMA3*, *IRT*, and *LCD* were analyzed by qPCR. Data are mean $\pm$ SD (n = 3).

3. Discussion

The presence of molecular hydrogen in plants has long been discovered [37–39]. In 2003, Dong and his colleagues discovered that exposing soils to hydrogen gas not only enhanced soil fertility, but also promoted plant growth of legumes and non-leguminous crops in both greenhouse and field trials. Afterwards, H_2, at first glance a simple molecule consisting of only two hydrogen atoms, was progressively suggested to be "a global player" in plant physiology [12,13,40], especially in laboratory levels [11,16].

For rice, previous greenhouse experiments illustrated that the application of metallic magnesium-produced HRW could influence the reproductive fitness of rice in greenhouse experiments, and in particular, significantly inhibited the thousand seed weight of conventional rice and Bt-transgenic rice [41]. Pot-based experiments further showed that conventional electrolytically produced HRW could enhance rice tolerance against salinity [42] and heavy metal stress (Cd and lead) [43]. However, the direct evidence of H_2 functioning in rice field trials, especially the influence in the field and grain traits, is still lacking. In this report, combined with nano-bubble technology, we provided physiological and molecular evidence for a previously uncharacterized role for H_2 positive control of rice field and grain quality traits in a trial experiment. Our results are significant for both fundamental and applied plant biology. The above conclusion was based on the following evidence.

First, the seed size of rice was an important issue in developmental biology, and was also an important part of seed yield [44,45]. A field study discovered the significant promotion of crop yield in spring wheat and barley when they were grown in H_2-treated soil [46]. Consistently, our results clearly showed that unlike the changes in length-width ratio (Figures 1E and 4D), the length, width, thickness, and thousand-seed weight of rice grain (Figure 1) and white rice (Figure 4) were remarkably improved by irrigating with HNW. The above results also suggested that compared to the length, the changes in the width of grain and the white rice are more sensitive to H_2.

These findings are inconsistent with those reported by Liu et al., in which they discovered the remarkable reduction in rice seed size in response to metallic magnesium-produced HRW [41]. The discrepancies may be attributed to different preparation methods for the hydrogen-based solution. Importantly, the possible negative influence achieved by other magnesium metabolites and/or possible pH alteration in the metallic magnesium-produced HRW could not be easily ruled out. A similar disadvantage was recently reported when magnesium hydride (MgH_2) was used as a preservative for prolonging the vase life of cut flowers [22].

The results of qPCR analysis (Figure 2) further indicated that the abovementioned HNW governing seed size might be achieved by regulating the expression level of specific genes in young panicles during the filling stage that controls rice seed size [47]. It is well documented that the final size of rice grains is coordinately controlled by cell proliferation and cell expansion in the spikelet hull [48,49]. Interestingly, the transcript levels of several typical genes controlling seed size was remarkably modulated by H_2. These include up-regulation of *RGB1* responsible for cellular proliferation [25], *SMG1* for grain length and width [26], *GS5* for grain width [27], and *GW8* for grain width and weight [29]. Previous results revealed that *SMG1*, which encodes mitogen-activated protein kinase kinase 4 (OsMKK4), could influence the seed size via influencing BR responses and the expression of BR-related genes [26]. *GS5* could keep BZR1/BRI1-associated receptor kinase1-7 (OsBAK1-7) on the cell surface, where it could interact with BZR1/BRI1 and enhance BR signaling, thereby affecting the grain size [27]. Meanwhile, the down-regulation of negatively correlated genes *GS3* for grain length and weight [29] were also consistent with the increased rice size in HNW-irrigated group.

Ample evidence shows that genes controlling hormone levels (BRs, etc.) could be used to increase grain yields in rice [49]. For instance, seeds of the BRs receptor and metabolism mutant were smaller than wild-type [27]. By contrast, seeds of transgenic Arabidopsis lines overexpressing BRs synthetic genes were larger than wild-type seeds [50]. Previous studies showed that H_2 might affect the synthesis and signaling of phytohormones, including auxin, abscisic acid, gibberellin, and ethylene [11,16]. Here, we observed that the transcripts of genes responsible for BRs synthesis/signaling (especially *CYP90D2*, *BZR1*, and *CPD*; Figure 3) and BRs-dependent seed size (*SMG1* [26]; *GS5* [31]; *NR1* [36]; and *AMT1* [33]) could be modulated by HNW (Figures 2 and 3), all of which were consistent with the increased seed size of rice observed after harvesting and processing (Figures 1 and 4). Thus, a cause-effect relationship between H_2 and BRs governing rice grain size should be carefully investigated in the near future.

It is well known that N, P, and K are principal nutrients that control crop productivity [51,52]. Therefore, the absorption and utilization efficiency of these nutrient elements are very important for rice yield [49,52]. To better understand the effects of HNW on crop production, the levels in P and K contents were evaluated. As expected, the irrigation with HNW could remarkably increase the accumulation of two elements in white rice (Figure 8C,D). Similar results were found in the positive changes in Mg and Fe ions (Figure 8E,F), both of which might be beneficial for human health in poor families, especially where rice is a staple food [53]. The reported results were also parallel to the transcriptional profiles of N, P, and K assimilation or the transport related gene (Figure 3), especially changes in *NRT2.3*, *NiR*, *ARE1*, *NLP4*, and *AKT1* transcripts, indicating that HNW control seed growth might be mediated by enhancing the assimilation of nutrient elements.

With the development of agriculture and food science and technology, the production for food crops also requires that agricultural products could improve qualitative characters and nutrient values [54]. Ample evidence confirmed that the increased gel consistency as well as decreasing protein and amylose contents in rice contributed greatly to the eating quality [4,54]. Our subsequent results shown in Figure 5 hinted that the white rice after irrigating with HNW had a better qualitative character, in terms of the changes in the above parameters, especially decreased amylose content. The latter change was supported by the transcriptional profiles of genes control of low amylose content (Figure 6).

Human health is inseparable from delicious, nutrient-rich, and, especially, safe foods [55]. Rice is a plant that might accumulate Cd in the grain. The production of rice around the world is challenged by Cd pollution and the subsequent elevation of grain Cd levels [56]. Previous hydroponic experiments revealed that HRW could reduce the absorption and accumulation of Cd in alfalfa seedlings [28,57]. Here, the results of our field trials showed that the application with HNW could remarkably decrease Cd accumulation by 52% in white rice (Figure 8A). This is a new finding, although recent results confirmed that HRW could respectively decease Cd and lead (Pb) contents in shoot or root tissues of rice (Xiangyaxiangzhan cultivar; [43]). These findings also suggest that HNW irrigation might not only help rice against Cd stress, but also reduce the accumulation of Cd in white rice and its processed products. The reduction in Cd content in white rice might be related to the changes in transcription levels of genes in roots at the filling stage related to metal transport (Figure 9). These include genes governing the entry of Cd in plants, such as *Nramp5* responsible for regulating the transport of Cd into the vascular bundle [35], *HMA2* and *HMA3* responsible for loading of Cd in the xylem as well as transporting Cd from the cytoplasm to the vacuole [3,35], *IRT1* for Cd entry to plants [3], and *LCD* responsible for regulating the phloem Cd transport [36]. We also noticed that although gene expression of *Nramp5* and *IRT1* (Figure 9) favors competition between Fe and Cd uptake [58], higher content of Fe and reduction in Cd accumulation in white rice were observed after HNW irrigation (Figure 8F). These results reflect the complexity of molecular hydrogen functions in plants growth and tolerance against heavy metals.

4. Materials and Methods

4.1. Plant Materials and Field Experiments

The rice (*Oryza sativa* L., Huruan1212 [59], a soft rice cultivar sensitive to sheath blight and rice blast; China Rice Data Center, https://www.ricedata.com, accessed on 9 October 2021) seeds were obtained from Zhenjiang Agricultural Research Institute, China. For further analysis, thirty-day-old rice seedlings were transplanted into paddy fields in Jurong,

Jiangsu province, China (longitude 119.26° E and latitude 31.95° N), in early June 2020, and allowed to grow in natural conditions. There were two paddies used for HNW and ditch water (Con) treatment groups, and every paddy was about 150 m^2 in this trial experiment. Additionally, the rice plants were grown without any chemical fertilizers and pesticides in the whole growth season. The daily highest and lowest temperature during the planting was recorded (Figure 10).

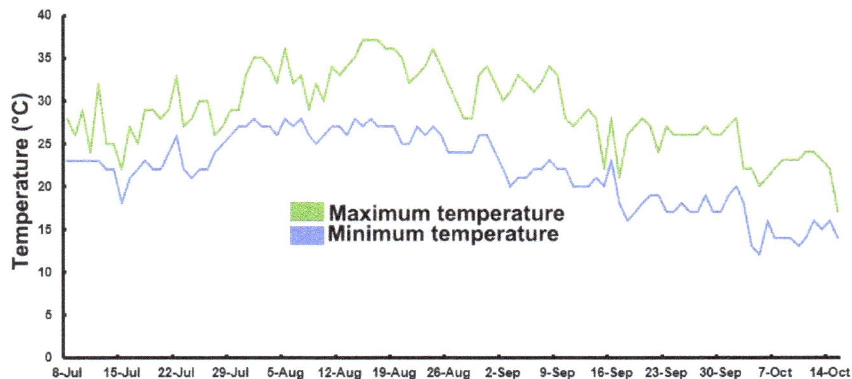

Figure 10. During rice planting, the field daily maximum and minimum temperature.

Rice plants were grown under the normal growing conditions until booting stage, and then HNW irrigation was imposed between 14 August 2020 and 15 October 2020, with one time per week, until the harvest stages. Each paddy field was irrigated with about 3 tons of water (HNW or ditch water) per time.

The HNW was produced by a hydrogen nanobubble water generator (HIM-22, Guangdong Cavolo Health Technology, China). Generally, when preparing HNW by electrolysis system, the voltage used for electrolysis was higher than 4.5 volts, and the current was higher than 35 amperes. H_2 produced from water electrolysis was infused into nanobubbles by a nanobubble aerator, and then dissolved into the ditch water. Before irrigating, the concentrations of dissolved H_2 were determined by a portable dissolved hydrogen meter (ENH-2000, TRUSTLEX, Japan; calibrated by gas chromatography). In our experimental conditions, H_2 content in the HNW was about 0.5 mM (1000 ppb). The half-time of dissolved H_2 in the above HNW was at least 3 h. Additionally, the diameter of the nanobubbles of hydrogen gas in the HNW was about 60–550 nm (determined by the NS300, Malvern Panalytical, Britain).

The rice was harvested on 1 November 2020, and the grains were then photographed and recorded. Afterwards, the grains were processed into white rice for further analysis. At least 1000 grains/white rice were randomly selected to record the sizes. Seed setting ratio and thousand seed weight were measured in triplicate. Each replicate included 1000 grains/white rice, and the total grains/white rice was 3000 (1000 × 3). Finally, about 15 g white rice was randomly selected and ground into powder for the further analysis.

4.2. Determination of Qualitative Characters

The content of total protein was measured by an automatic kieldahl apparatus (KDN-08A; Hongji, Shanghai, China) according to the previous study [60].

The amylose contents were determined using a dual-wavelength iodine-binding method [61].

The determination of gel consistency, chalky rice rate, and starch content were carried out according to the previous method [4].

4.3. Determination of Ion Content

According to previous reports [62], ion contents in white rice and soil were detected by using a Digital Block Sample Digestion System (LabTech ED54 DigiBlock). The metal ions contents were determined by an Inductively Coupled Plasma-Optical Emission Spectrometer (ICP-OES, iCAP 7000, Thermo Fisher).

4.4. Determination of Albumin, Globulin, Prolamin, and Glutelin Content

After extraction [63], the composition of white rice protein, including albumin, globulin, prolamin, and glutelin, was determined by using BCA Protein Assay Kit (TaKaRa, Beijing, China).

4.5. Real-Time Quantitative Reverse Transcription-PCR (qPCR)

The leaves, young panicles, and roots were randomly collected from rice plants in the filling stage (20 September 2020), and further frozen in liquid nitrogen immediately. After the extraction of total RNA and the synthesis of cDNA, the quantitative real-time PCR (qPCR) was carried out. The primers' sequences are shown in Supplementary Table S1. The relative expression levels of corresponding genes are presented as values relative to those of corresponding control samples, after normalization with two reference genes *OsActin1* and *OsUbi*. The results of relative genes expression levels were analyzed by the $2^{-\Delta\Delta CT}$ method [64]. All determinations were carried out using three separate RNA samples, and each run in triplicate.

4.6. Statistical Analysis

All results are shown as the mean values $\pm$ SD of three independent experiments with three biological replicates for each. By using Origin 2021, the data were analyzed by Student's *t*-test or one-way analysis of variance (ANOVA). $p < 0.05$ was considered statistically significant.

5. Conclusions

In conclusion, this study demonstrated that the application of HNW during the growth and development stage of rice could not only significantly increase the field and grain quantity of rice grains and white rice, but could also improve qualitative characters, maintain nutrition ingredients, and alleviate accumulation of Cd of the white rice. As we know, this is the first time using HNW in field trails, which might act as an important strategy for enhancing sustainable crop production. Importantly, results on the transcriptional profiling seem to give some answers on the question of how HNW increased field and grain quality traits. Since a multitude of different signaling pathways could be closely associated with molecular hydrogen functions in plants [16], the existence of a simple cause-and-effect chain seems very unlikely.

Compared to the direct usage of hydrogen gas in field soil [46], irrigation with HNW is a relatively convenient and cheaper approach; thus, this method might be used in large-scale agriculture. We also admitted that hydrogen-based agriculture is just at the beginning stage, and many field trials and deeper molecular mechanisms should be further carried out and elucidated.

Supplementary Materials: The following are available online at https://www.mdpi.com/2223-7747/10/11/2331/s1, Table S1: The sequences of primers for qPCR, References [65–75] are cited in the Supplementary Materials.

Author Contributions: P.C. and W.S. designed and refined the research; P.C., L.K., D.J., J.Y. and Z.S. performed research; P.C., J.W., W.L., T.Z., L.H., W.Z. and Q.Y. analyzed data; P.C. and W.S. wrote the article. P.C., Z.Z. and W.S. revised the article. All authors discussed the results and comments on the manuscript. All authors have read and agreed to the published version of the manuscript.

Funding: This work was financially supported by the funding from Center of Hydrogen Science, 412 Shanghai Jiao Tong University (China), Foshan Agriculture Science and Technology Project (Fo-413 shan City Budget No. 140, 2019.), and Air Liquide (China) R&D Co., Ltd.

Data Availability Statement: This statement if the study did not report any data.

Acknowledgments: This work was financially supported by the Funding from Center of Hydrogen Science, Shanghai Jiao Tong University (China), Foshan Agriculture Science and Technology Project (Foshan City Budget No. 140, 2019.), and Air Liquide (China) R&D Co., Ltd.

Conflicts of Interest: The authors declare no conflict of interest.

References

1. Law, J.W.-F.; Ser, H.-L.; Khan, T.M.; Chuah, L.-H.; Pusparajah, P.; Chan, K.-G.; Goh, B.-H.; Lee, L.-H. The Potential of *Streptomyces* as Biocontrol Agents against the Rice Blast Fungus, Magnaporthe oryzae (Pyricularia oryzae). *Front. Microbiol.* **2017.** *8*, 3. [CrossRef] [PubMed]
2. Chen, H.; Siebenmorgen, T.J.; Griffin, K. Quality Characteristics of Long-Grain Rice Milled in Two Commercial Systems. *Cereal Chem. J.* **1998**, *75*, 560–565. [CrossRef]
3. Ueno, D.; Yamaji, N.; Kono, I.; Huang, C.F.; Ando, T.; Yano, M.; Ma, J.F. Gene limiting cadmium accumulation in rice. *Proc. Natl. Acad. Sci. USA* **2010**, *107*, 16500–16505. [CrossRef]
4. Zhang, C.; Chen, S.; Ren, X.; Lu, Y.; Liu, D.; Cai, X.; Li, Q.; Gao, J.; Liu, Q. Molecular Structure and Physicochemical Properties of Starches from Rice with Different Amylose Contents Resulting from Modification of OsGBSSI Activity. *J. Agric. Food Chem.* **2017**, *65*, 2222–2232. [CrossRef] [PubMed]
5. Soltis, P.S.; Soltis, D.E. The Role of Hybridization in Plant Speciation. *Annu. Rev. Plant Biol.* **2009**, *60*, 561–588. [CrossRef] [PubMed]
6. Qian, L.; Zhang, C.; Zuo, F.; Zheng, L.; Li, D.; Zhang, A.; Zhang, D. Effects of fertilizers and pesticides on the mineral elements used for the geographical origin traceability of rice. *J. Food Compos. Anal.* **2019**, *83*, 103276. [CrossRef]
7. Moose, S.P.; Mumm, R.H. Molecular Plant Breeding as the Foundation for 21st Century Crop Improvement. *Plant Physiol.* **2008**, *147*, 969–977. [CrossRef]
8. Yu, Q.; Liu, S.; Yu, L.; Xiao, Y.; Zhang, S.; Wang, X.; Xu, Y.; Yu, H.; Li, Y.; Yang, J.; et al. RNA demethylation increases the yield and biomass of rice and potato plants in field trials. *Nat. Biotechnol.* **2021**. [CrossRef]
9. Jiang, L. Commercialization of the gene-edited crop and morality: Challenges from the liberal patent law and the strict GMO law in the EU. *New Genet. Soc.* **2019**, *39*, 191–218. [CrossRef]
10. Ohsawa, I.; Ishikawa, M.; Takahashi, K.; Watanabe, M.; Nishimaki, K.; Yamagata, K.; Katsura, K.-I.; Katayama, Y.; Asoh, S.; Ohta, S. Hydrogen acts as a therapeutic antioxidant by selectively reducing cytotoxic oxygen radicals. *Nat. Med.* **2007**, *13*, 688–694. [CrossRef] [PubMed]
11. Shen, W.; Sun, X. Hydrogen biology: It is just beginning. *Chin. J. Biochem. Mol. Bio.* **2019**, *35*, 1037–1050.
12. Jin, Q.; Zhu, K.; Cui, W.; Xie, Y.; Han, B.; Shen, W. Hydrogen gas acts as a novel bioactive molecule in enhancing plant tolerance to paraquat-induced oxidative stress via the modulation of heme oxygenase-1 signalling system. *Plant Cell Environ.* **2012**, *36*, 956–969. [CrossRef] [PubMed]
13. Xie, Y.; Mao, Y.; Lai, D.; Zhang, W.; Shen, W. H2 enhances Arabidopsis salt tolerance by manipulating ZAT10/12-mediated anti-oxidant defence and controlling sodium exclusion. *PLoS ONE* **2012**, *7*, e49800. [CrossRef]
14. Xie, Y.; Mao, Y.; Zhang, W.; Lai, D.; Wang, Q.; Shen, W. Reactive Oxygen Species-Dependent Nitric Oxide Production Contributes to Hydrogen-Promoted Stomatal Closure in Arabidopsis. *Plant Physiol.* **2014**, *165*, 759–773. [CrossRef] [PubMed]
15. Cao, Z.; Duan, X.; Yao, P.; Cui, W.; Cheng, D.; Zhang, J.; Jin, Q.; Chen, J.; Dai, T.; Shen, W. Hydrogen Gas Is Involved in Auxin-Induced Lateral Root Formation by Modulating Nitric Oxide Synthesis. *Int. J. Mol. Sci.* **2017**, *18*, 2084. [CrossRef] [PubMed]
16. Li, L.; Lou, W.; Kong, L.; Shen, W. Hydrogen Commonly Applicable from Medicine to Agriculture: From Molecular Mechanisms to the Field. *Curr. Pharm. Des.* **2021**, *27*, 747–759. [CrossRef]
17. Hu, H.; Zhao, S.; Li, P.; Shen, W. Hydrogen gas prolongs the shelf life of kiwifruit by decreasing ethylene biosynthesis. *Postharvest Biol. Technol.* **2018**, *135*, 123–130. [CrossRef]
18. Hu, H.; Li, P.; Shen, W. Preharvest application of hydrogen-rich water not only affects daylily bud yield but also contributes to the alleviation of bud browning. *Sci. Hortic.* **2021**, *287*, 110267. [CrossRef]
19. Su, J.; Nie, Y.; Zhao, G.; Cheng, D.; Wang, R.; Chen, J.; Zhang, S.; Shen, W. Endogenous hydrogen gas delays petal senescence and extends the vase life of lisianthus cut flowers. *Postharvest Biol. Technol.* **2019**, *147*, 148–155. [CrossRef]
20. Li, L.; Liu, Y.; Wang, S.; Zou, J.; Ding, W.; Shen, W. Magnesium Hydride-Mediated Sustainable Hydrogen Supply Prolongs the Vase Life of Cut Carnation Flowers via Hydrogen Sulfide. *Front. Plant Sci.* **2020**, *11*, 595376. [CrossRef]
21. Jiang, K.; Kuang, Y.; Feng, L.; Liu, Y.; Wang, S.; Du, H.; Shen, W. Molecular Hydrogen Maintains the Storage Quality of Chinese Chive through Improving Antioxidant Capacity. *Plants* **2021**, *10*, 1095. [CrossRef] [PubMed]

22. Nghia, N.H.; van Giang, P.T.; Hanh, N.T.; St-Hilaire, S.; Domingos, A.J. Control of Vibrioparahaemolyticus (AHPND strain) and improvement of water quality using nanobubble technology. *Aquac. Res.* **2021**, *52*, 2727–2739. [CrossRef]

23. Agarwal, A.; Ng, W.J.; Liu, Y. Principle and applications of microbubble and nanobubble technology for water treatment. *Chemosphere* **2011**, *84*, 1175–1180. [CrossRef]

24. Fan, W.; Zhang, Y.; Liu, S.; Li, X.; Li, J. Alleviation of copper toxicity in *Daphnia magna* by hydrogen nanobubble water. *J. Hazard. Mater.* **2020**, *389*, 122155. [CrossRef] [PubMed]

25. Utsunomiya, Y.; Samejima, C.; Takayanagi, Y.; Izawa, Y.; Yoshida, T.; Sawada, Y.; Fujisawa, Y.; Kato, H.; Iwasaki, Y. Suppression of the rice heterotrimeric G protein β-subunit gene, RGB1, causes dwarfism and browning of internodes and lamina joint regions. *Plant J.* **2011**, *67*, 907–916. [CrossRef] [PubMed]

26. Duan, P.; Rao, Y.; Zeng, D.; Yang, Y.; Xu, R.; Zhang, B.; Dong, G.; Qian, Q.; Li, Y. SMALL GRAIN 1, which encodes a mitogen-activated protein kinase kinase 4, influences grain size in rice. *Plant J.* **2014**, *77*, 547–557. [CrossRef] [PubMed]

27. Li, Y.; Fan, C.; Xing, Y.; Jiang, Y.; Luo, L.; Sun, L.; Shao, D.; Xu, C.; Li, X.; Xiao, J.; et al. Natural variation in GS5 plays an important role in regulating grain size and yield in rice. *Nat. Genet.* **2011**, *43*, 1266–1269. [CrossRef] [PubMed]

28. Zhang, X.; Wang, J.; Huang, J.; Lan, H.; Wang, C.; Yin, C.; Wu, Y.; Tang, H.; Qian, Q.; Li, J.; et al. Rare allele of OsPPKL1 associated with grain length causes extra-large grain and a significant yield increase in rice. *Proc. Natl. Acad. Sci.* **2012**, *109*, 21534–21539. [CrossRef] [PubMed]

29. Xia, D.; Zhou, H.; Liu, R.; Dan, W.; Li, P.; Wu, B.; Chen, J.; Wang, L.; Gao, G.; Zhang, Q.; et al. GL3.3, a Novel QTL Encoding a GSK3/SHAGGY-like Kinase, Epistatically Interacts with GS3 to Produce Extra-long Grains in Rice. *Mol. Plant* **2018**, *11*, 754–756. [CrossRef]

30. Han, X.; Wu, K.; Fu, X.; Liu, Q. Improving coordination of plant growth and nitrogen metabolism for sustainable agriculture. *aBIOTECH* **2020**, *1*, 255–275. [CrossRef]

31. Mora-García, S.; Vert, G.; Yin, Y.; Caño-Delgado, A.; Cheong, H.; Chory, J. Nuclear protein phosphatases with Kelch-repeat domains modulate the response to brassinosteroids in Arabidopsis. *Genes Dev.* **2004**, *18*, 448–460. [CrossRef]

32. Das, T.; Shukla, Y.M.; Poonia, T.C.; Meena, M.; Meena, M.D. Effects of brassinolide on physiological characteristics of rice (*Oryza sativa* L.) with different salinity levels. *Ann. Biol.* **2013**, *29*, 228–231.

33. Yang, S.; Yuan, D.; Zhang, Y.; Sun, Q.; Xuan, Y.H. BZR1 Regulates Brassinosteroid-Mediated Activation of AMT1;2 in Rice. *Front. Plant Sci.* **2021**, *12*, 665883. [CrossRef] [PubMed]

34. Rayee, R.; Xuan, T.; Khanh, T.; Tran, H.-D.; Kifayatullah, K. Efficacy of Irrigation Interval after Anthesis on Grain Quality, Alkali Digestion, and Gel Consistency of Rice. *Agriculture* **2021**, *11*, 325. [CrossRef]

35. Sasaki, A.; Yamaji, N.; Yokosho, K.; Ma, J.F. Nramp5 Is a Major Transporter Responsible for Manganese and Cadmium Uptake in Rice. *Plant Cell* **2012**, *24*, 2155–2167. [CrossRef]

36. Shimo, H.; Ishimaru, Y.; An, G.; Yamakawa, T.; Nakanishi, H.; Nishizawa, N.K. Low cadmium (LCD), a novel gene related to cadmium tolerance and accumulation in rice. *J. Exp. Bot.* **2011**, *62*, 5727–5734. [CrossRef] [PubMed]

37. Gaffron, H. Reduction of Carbon Dioxide with Molecular Hydrogen in Green Algæ. *Nat. Cell Biol.* **1939**, *143*, 204–205. [CrossRef]

38. Gest, H.; Kamen, M.D. Photoproduction of Molecular Hydrogen by Rhodospirillum rubrum. *Science* **1949**, *109*, 558–559. [CrossRef]

39. Renwick, G.M.; Giumarro, C.; Siegel, S.M. Hydrogen Metabolism in Higher Plants. *Plant Physiol.* **1964**, *39*, 303–306. [CrossRef]

40. Zeng, J.; Zhang, M.; Sun, X. Molecular Hydrogen Is Involved in Phytohormone Signaling and Stress Responses in Plants. *PLOS ONE* **2013**, *8*, e71038. [CrossRef] [PubMed]

41. Liu, F.; Jiang, W.; Han, W.; Li, J.; Liu, Y. Effects of Hydrogen-Rich Water on Fitness Parameters of Rice Plants. *Agron. J.* **2017**, *109*, 2033–2039. [CrossRef]

42. Fu, X.; Ma, L.; Gui, R.; Ashraf, U.; Li, Y.; Yang, X.; Zhang, J.; Imran, M.; Tang, X.; Tian, H.; et al. Differential response of fragrant rice cultivars to salinity and hydrogen rich water in relation to growth and antioxidative defense mechanisms. *Int. J. Phytoremediation* **2021**, *23*, 1203–1211. [CrossRef] [PubMed]

43. Ma, L.; Kong, L.; Gui, R.; Yang, X.; Zhang, J.; Gong, Q.; Qin, D.; Zhuang, M.; Ashraf, U.; Mo, Z. Application of hydrogen-rich water modulates physio-biochemical functions and early growth of fragrant rice under Cd and Pb stress. *Environ. Sci. Pollut. Res.* **2021**, *28*, 58558–58569. [CrossRef] [PubMed]

44. Moles, A.T.; Ackerly, D.D.; Webb, C.O.; Tweddle, J.C.; Dickie, J.B.; Pitman, A.J.; Westoby, M. Factors that shape seed mass evolution. *Proc. Natl. Acad. Sci.* **2005**, *102*, 10540–10544. [CrossRef] [PubMed]

45. Radny, J.; van der Putten, W.H.; Tielbörger, K.; Meyer, K.M. Influence of seed size on performance of non-native annual plant species in a novel community at two planting densities. *Acta Oecologica* **2018**, *92*, 131–137. [CrossRef]

46. Dong, Z.; Wu, L.; Kettlewell, B.; Caldwell, C.D.; Layzell, D.B. Hydrogen fertilization of soils—Is this a benefit of legumes in rotation? *Plant Cell Environ.* **2003**, *26*, 1875–1879. [CrossRef]

47. Zuo, J.; Li, J. Molecular Genetic Dissection of Quantitative Trait Loci Regulating Rice Grain Size. *Annu. Rev. Genet.* **2014**, *48*, 99–118. [CrossRef] [PubMed]

48. Li, N.; Xu, R.; Duan, P.; Li, Y. Control of grain size in rice. *Plant Reprod.* **2018**, *31*, 237–251. [CrossRef] [PubMed]

49. Li, N.; Xu, R.; Li, Y. Molecular Networks of Seed Size Control in Plants. *Annu. Rev. Plant Biol.* **2019**, *70*, 435–463. [CrossRef]

50. Xu, C.; Liu, Y.; Li, Y.; Xu, X.; Xu, C.; Li, X.; Xiao, J.; Zhang, Q. Differential expression of GS5 regulates grain size in rice. *J. Exp. Bot.* **2015**, *66*, 2611–2623. [CrossRef] [PubMed]

51. Oldroyd, G.E.D.; Leyser, O. A plant's diet, surviving in a variable nutrient environment. *Science* **2020**, *368*, 0196. [CrossRef]

52. Wang, Y.; Chen, Y.; Wu, W. Potassium and phosphorus transport and signaling in plants. *J. Integr. Plant Biol.* **2021**, *63*, 34–52. [CrossRef]

53. Balakrishna, A.K.; Auckaili, A.; Farid, M. Effect of high pressure impregnation on micronutrient transfer in rice. *Food Chem.* **2021**, *362*, 130244. [CrossRef]

54. Van Dongen, M.V.; van den Berg, M.C.; Vink, N.; Kok, F.J.; de Graaf, C. Taste-nutrient relationships in commonly consumed foods. *Br. J. Nutr.* **2012**, *108*, 140–147. [CrossRef] [PubMed]

55. Pogoson, E.; Carey, M.; Meharg, C.; Meharg, A. Reducing the cadmium, inorganic arsenic and dimethylarsinic acid content of rice through food-safe chemical cooking pre-treatment. *Food Chem.* **2021**, *338*, 127842. [CrossRef] [PubMed]

56. Sebastian, A.; Prasad, M.N.V. Cadmium minimization in rice. A review. *Agron. Sustain. Dev.* **2014**, *34*, 155–173. [CrossRef]

57. Dai, C.; Cui, W.; Pan, J.; Xie, Y.; Wang, J.; Shen, W. Proteomic analysis provides insights into the molecular bases of hydrogen gas-induced cadmium resistance in Medicago sativa. *J. Proteom.* **2017**, *152*, 109–120. [CrossRef]

58. He, X.L.; Fan, S.K.; Zhu, J.; Guan, M.Y.; Liu, X.X.; Zhang, Y.S.; Jin, C.W. Iron supply prevents Cd uptake in Arabidopsis by inhibiting IRT1 expression and favoring competition between Fe and Cd uptake. *Plant Soil* **2017**, *416*, 453–462. [CrossRef]

59. Chen, Y.; Wang, M.; Ouwerkerk, P.B.F. Molecular and environmental factors determining grain quality in rice. *Food Energy Secur.* **2012**, *1*, 111–132. [CrossRef]

60. Suo, B.; Li, H.; Wang, Y.; Li, Z.; Pan, Z.; Ai, Z. Effects of ZnO nanoparticle-coated packagingfilm on pork meat quality during cold storage. *J. Sci. Food Agric.* **2017**, *97*, 2023–2029. [CrossRef]

61. Zhu, T.; Jackson, D.S.; Wehling, R.L.; Geera, B. Comparison of Amylose Determination Methods and the Development of a Dual Wavelength Iodine Binding Technique. *Cereal Chem. J.* **2008**, *85*, 51–58. [CrossRef]

62. Cui, W.; Gao, C.; Fang, P.; Lin, G.; Shen, W. Alleviation of cadmium toxicity in Medicago sativa by hydrogen-rich water. *J. Hazard. Mater.* **2013**, *260*, 715–724. [CrossRef] [PubMed]

63. Chen, P.; Shen, Z.; Ming, L.; Li, Y.; Dan, W.; Lou, G.; Peng, B.; Wu, B.; Li, Y.; Zhao, D.; et al. Genetic Basis of Variation in Rice Seed Storage Protein (Albumin, Globulin, Prolamin, and Glutelin) Content Revealed by Genome-Wide Association Analysis. *Front. Plant Sci.* **2018**, *9*, 612. [CrossRef] [PubMed]

64. Livak, K.J.; Schmittgen, T.D. Analysis of Relative Gene Expression Data Using Real-Time Quantitative PCR and the $2^{-\Delta\Delta C}{}_{T}$ Method. *Methods* **2001**, *25*, 402–408. [CrossRef]

65. Yang, B.; Wang, X.M.; Ma, H.Y.; Li, Y.J.; Dai, C.C. Effects of the fungal endophyte Phomopsis liquidambari on nitrogen uptake and me-tabolism in rice. *Plant Growth Regul.* **2014**, *73*, 165–179.

66. Yan, M.; Fan, X.; Feng, H.; Miller, A.J.; Shen, Q.; Xu, G. Rice OsNAR2.1 interacts with OsNRT2.1, OsNRT2.2 and OsNRT2.3a nitrate transporters to provide uptake over high and low concentration ranges. *Plant Cell Environ.* **2011**, *34*, 1360–1372. [CrossRef]

67. Wang, Q.; Su, Q.; Nian, J.; Zhang, J.; Guo, M.; Dong, G.; Hu, J.; Wang, R.; Wei, C.; Li, G.; et al. The Ghd7 transcription factor represses ARE1 expression to enhance nitrogen utilization and grain yield in rice. *Mol. Plant* **2021**, *14*, 1012–1023. [CrossRef]

68. Ye, Y.; Li, P.; Xu, T.; Zeng, L.; Cheng, D.; Yang, M.; Luo, J.; Lian, X. OsPT4 Contributes to Arsenate Uptake and Transport in Rice. *Front. Plant Sci.* **2017**, *8*, 2197. [CrossRef]

69. Okada, T.; Nakayama, H.; Shinmyo, A.; Yoshida, K. Expression of OsHAK genes encoding potassium ion transporters in rice. *Plant Biotechnol.* **2008**, *25*, 241–245. [CrossRef]

70. Sakamoto, T.; Matsuoka, M. Characterization of CONSTITUTIVE PHOTOMORPHOGENESIS AND DWARFISM Homologs in Rice (Oryza sativa L.). *J. Plant Growth Regul.* **2006**, *25*, 245–251. [CrossRef]

71. Tsukagoshi, Y.; Suzuki, H.; Seki, H.; Muranaka, T.; Ohyama, K.; Fujimoto, Y. Ajuga Δ24-Sterol Reductase Catalyzes the Direct Reductive Conversion of 24-Methylenecholesterol to Campesterol. *J. Biol. Chem.* **2016**, *291*, 8189–8198. [CrossRef]

72. Crofts, N.; Satoh, Y.; Miura, S.; Hosaka, Y.; Abe, M.; Fujita, N. Active-type starch synthase (SS) IIa from indica rice partially com-plements the sugary-1 phenotype in japonica rice endosperm. *Plant Mol. Biol.* **2021**. [CrossRef]

73. Utsumi, Y.; Nakamura, Y. Structural and enzymatic characterization of the isoamylase1 homo-oligomer and the isoamylase1–isoamylase2 hetero-oligomer from rice endosperm. *Planta* **2006**, *225*, 75–87. [CrossRef]

74. Satoh, H.; Nishi, A.; Yamashita, K.; Takemoto, Y.; Tanaka, Y.; Hosaka, Y.; Sakurai, A.; Fujita, N.; Nakamura, Y. Starch-branching enzyme I-deficient mutation spe-cifically affects the structure and properties of starch in rice endosperm. *Plant Physiol.* **2003**, *133*, 1111–1121.

75. Nakamura, Y.; Utsumi, Y.; Sawada, T.; Aihara, S.; Utsumi, C.; Yoshida, M.; Kitamura, S. Characterization of the reactions of starch branching enzymes from rice endosperm. *Plant Cell Physiol.* **2010**, *51*, 776–794. [CrossRef]

Article

The Involvement of Glucose in Hydrogen Gas-Medicated Adventitious Rooting in Cucumber

Zongxi Zhao, Changxia Li, Huwei Liu, Jingjing Yang, Panpan Huang and Weibiao Liao *

College of Horticulture, Gansu Agricultural University, 1 Yinmen Village, Anning District, Lanzhou 730070, China; zhaozongxi2021@163.com (Z.Z.); licxgsau5@163.com (C.L.); lhwgsauedu@163.com (H.L.); yangjingjing202108@163.com (J.Y.); huangpp2243739575@163.com (P.H.)
* Correspondence: liaowb@gsau.edu.cn; Tel.: +86-931-7632399; Fax: +86-931-7632155

Abstract: Hydrogen gas (H_2) and glucose (Glc) have been reported as novel antioxidants and signal molecules involved in multiple biological processes in plants. However, the physiological roles and relationships of H_2 and Glc in adventitious rooting are less clear. Here, we showed that the effects of different concentrations Glc (0, 0.01, 0.05, 0.10, 0.50 and 1.00 mM) on adventitious rooting in cucumber were dose-dependent, with a maximal biological response at 0.10 mM. While, the positive roles of hydrogen rich water (HRW, a H_2 donor)-regulated adventitious rooting were blocked by a specific Glc inhibitor glucosamine (GlcN), suggesting that Glc might be responsible for H_2-regulated adventitious root development. HRW increased glucose, sucrose, starch and total sugar contents. Glucose-6-phosphate (G6P), fructose-6-phosphate (F6P) and glucose-1-phosphate (G1P) contents were also increased by HRW. Meanwhile, the activities of sucrose-related enzymes incorporating sucrose synthase (SS) and sucrose phosphate synthase (SPS) and glucose-related enzymes including hexokinase (HK), pyruvate kinase (PK) and adenosine 5′-diphosphate pyrophosphorylase (AGPase) were increased by HRW. Moreover, HRW upregulated the expression levels of sucrose or glucose metabolism-related genes including *CsSuSy1*, *CsSuSy6*, *CsHK1*, *CsHK3*, *CsUDP1*, *CsUDP1-like*, *CsG6P1* and *CsG6P1-like*. However, these positive roles were all inhibited by GlcN. Together, H_2 might regulate adventitious rooting by promoting glucose metabolism.

Keywords: glucosamine; sucrose; starch; gene expression; sugar metabolism

Citation: Zhao, Z.: Li, C.; Liu, H.; Yang, J.; Huang, P.; Liao, W. The Involvement of Glucose in Hydrogen Gas-Medicated Adventitious Rooting in Cucumber. *Plants* **2021**, *10*, 1937. https://doi.org/10.3390/plants10091937

Academic Editor: John Hancock

Received: 19 August 2021
Accepted: 14 September 2021
Published: 17 September 2021

Publisher's Note: MDPI stays neutral with regard to jurisdictional claims in published maps and institutional affiliations.

1. Introduction

Adventitious roots (AR) are postembryonic roots which originate from the stem, leaf petiole and non-pericycle tissue of old roots [1]. Normally, inappropriate conditions including injury and stress promote AR formation. Recent results revealed that AR formation was positively regulated by plant hormones and signaling molecules, such as abscisic acid (ABA) [2], auxin [3], brassinolide (BR) [4], gibberellin (GA) [5], ethylene [6], hydrogen peroxide (H_2O_2) [7], nitric oxide (NO) [8], carbon monoxide (CO) [9] and hydrogen sulfide (H_2S) [10] Additionally, molecular evidence illustrated that auxin- and ethylene-related genes and proteins are closely associated with the initiation and development of AR [11]. The complex responses regulated by these hormones, signaling molecules, genes and proteins are most likely to be achieved by a more detailed AR signaling process. However, whether there are some other novel singling molecule(s) involved in AR formation remains to be studied.

Hydrogen gas (H_2), colorless, tasteless and flammable, is the structurally simplest gas in nature. Recently, research on H_2 has progressed from focusing on its role as a fuel to its role as matter that is able to regulate multiple biological functions in animals and plants. Treatment with H_2 relieved brain damage and inflammation after traumatic brain injury in rats [12]. At the same time, it was reported that hydrogen therapy was a potential and effective treatment for exercise-induced injury in sports medicine [13]. In plants, H_2 is considered as an important signaling modulator that functions in plant responses against salt, heavy metals, low temperatures and paraquat stresses [14]. H_2 actively controls a series of plant growth and development stages including seed germination, seedling growth, adventitious rooting and root elongation [14]. In addition, hydrogen rich water (HRW) could defer postharvest ripening and aging of kiwifruit [15]. Interestingly, H_2 can interact directly with specific signaling pathways, including NO, CO and ethylene [16].

Starch is a ubiquitous storage polysaccharide in the plant kingdom. Adenosine $5'$-diphosphate pyrophosphorylase (AGPase) is a key enzyme governing starch synthesis. In the process of plant metabolism, starch is hydrolyzed into monosaccharides such as glucose (Glc). Glc is an important member of monosaccharides which exists widely in animals and plants. Simultaneously, Glc is recognized as a central signaling molecule that balances the requirement of nutrient and energy in plants. Hexokinase (HK), as an enzyme in glycolysis, phosphorylates glucose to glucose-6-phosphate [17]. Both HK-dependent and HK-independent glucose signal transduction pathways appear to coexist in plants [18]. Sucrose is synthesized using cytosolic phosphotriose as a substrate. Sucrose may then be transported to sink organs or it is cleaved by invertase to Glc and fructose [19]. Fructose-6-phosphate (F6P) and sucrose phosphate synthase (SPS) have been recognized as the rate-limiting products or enzymes in the process of sucrose synthesis [20]. Glc acts as a primary signal molecule in plant response against drought stress and heat stress [21]. Simultaneously, it was reported that Glc regulated a series of growth and development stages, including seed germination, seedling development, embryo development, cell division, stomatal movement, seed dormancy, leaf senescence and fruit ripening [22,23]. Previous results also elaborated that Glc significantly regulated AR development in *Arabidopsis thaliana* [24].

Previous studies have shown that both H_2 and Glc regulate AR development in plants. However, there is no information regarding the crosstalk between H_2 and Glc in adventitious rooting. The purpose of the research study was to investigate the roles and interaction of H_2 and Glc in AR development. Deeper insights into the interplay of various signaling molecules with H_2 will help provide a road-map for H_2-regulated AR development.

2. Results

2.1. Effect of Different Concentrations Glc on AR Development

When compared with the control, 0.01 mM Glc had no significant effect on root number (Table 1). However, treatments with 0.05, 0.10 and 0.50 mM Glc significantly increased root number (Table 1). Root length was higher in 0.01, 0.05, 0.10 and 0.50 mM Glc treatments than in the control (Table 1). Compared to the control, 1.00 mM Glc significantly inhibited both root number and root length. The results indicated that Glc treatment affected AR development in a dose-dependent manner. Among the different concentrations, the maximum root number and root length were observed at 0.10 mM concentration of Glc (Table 1). Thus, 0.10 mM Glc was used as a treatment for further studies during the rooting process.

Table 1. Effect of Glc on AR development in cucumber.

Glc/mM	Root Number	Root Length (mm)
0.00	3.21 ± 0.18 c	4.99 ± 0.18 d
0.01	3.91 ± 0.12 c	6.97 ± 0.44 c
0.05	4.63 ± 0.14 b	8.04 ± 0.13 b
0.10	6.95 ± 0.05 a	9.69 ± 005 a
0.50	4.37 ± 0.01 b	7.06 ± 0.25 c
1.00	2.58 ± 0.18 d	4.06 ± 0.44 e

Effect of different concentrations Glc (0, 0.01, 0.05, 0.10, 0.50 and 1.00 mM) on AR development in cucumber explants. The values [mean ± standard error (SE)] are the average of three independent experiments (n = 30 explants per replicate). Values not sharing the same letters in the same list were significantly different by Duncan's multiple-comparison test ($p < 0.05$).

2.2. Involvement of Glc in HRW-Regulated AR Development

Our previous study found that HRW treatment could promote the formation of AR in cucumber, and the optimal concentration was 50% [25], which were also used in this study. To analyze the roles of Glc in H_2-regulated AR development, cucumber explants were treated with 50% HRW, 0.10 mM Glc and 0.10 mM Glc inhibitor glucosamine (GlcN) alone or together (Figure 1). Cucumber explants treated with 50% HRW or 0.10 mM Glc exhibited significant increase in root number and root length (Figure 1). Additionally, co-treatment with HRW and Glc significantly enhanced adventitious rooting in comparison with HRW or Glc treatment alone. When GlcN was added to HRW, the positive effects of HRW on rooting was weakened. GlcN treatment alone significantly inhibited adventitious rooting in comparison with the control (Figure 1). These results indicated that Glc might be involved in H_2-promoted AR development in cucumber.

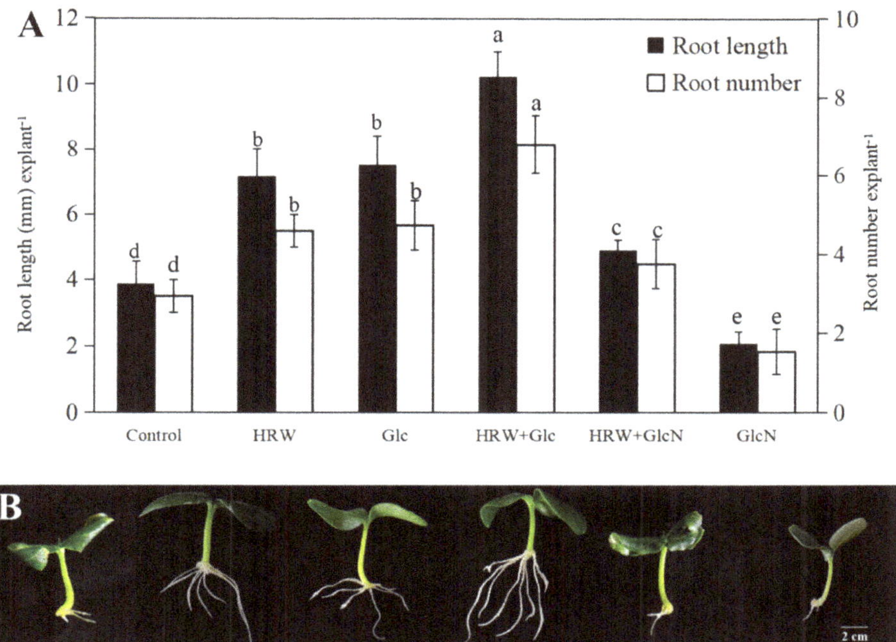

Figure 1. Effects of HRW, Glc and GlcN on root number and root length (**A**) and phenotype (**B**) of cucumber explants treated with 50% HRW, 0.10 mM Glc and 0.10 mM GlcN alone or together. n = 30 explants per replicate. Bars with different lower case letters were significantly different ($p < 0.5$). HRW: hydrogen rich water; Glc: glucose; GlcN: glucosamine.

2.3. Effects of HRW, Glc and GlcN on Glucose, Sucrose, Starch and Total Sugar Contents during Adventitious Rooting

Compared with the control, HRW and Glc treatments significantly increased the contents of glucose, sucrose, starch and total sugar (Figure 2). Additionally, glucose, sucrose, starch and total sugar contents in explants treated with HRW plus Glc were significantly increased compared to the HRW or Glc alone treatments. The glucose, sucrose, starch and total sugar contents in Glc inhibitor GlcN treatment were lower than that in the control (Figure 2).

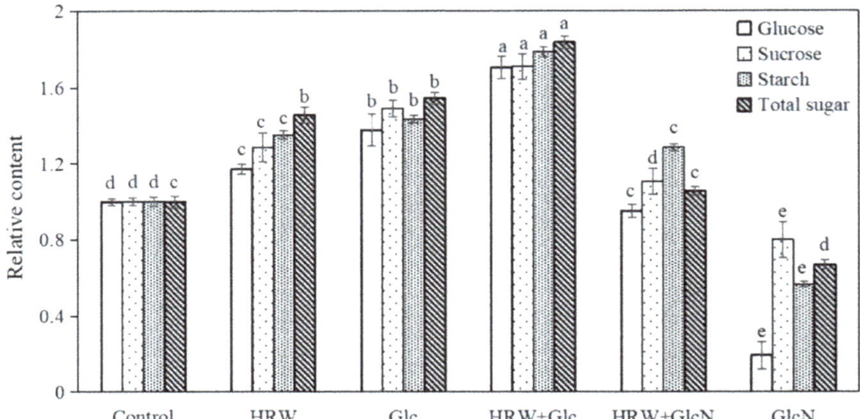

Figure 2. Effects of HRW, Glc and GlcN on glucose, sucrose, starch, and total sugar contents during adventitious rooting. n = 30 explants per replicate. Bars with different lower case letters were significantly different ($p < 0.5$). HRW: hydrogen rich water; Glc: glucose; GlcN glucosamine. The relative content (compared to the control) is displayed in the figure.

2.4. Effects of HRW, Glc and GlcN on Hexose Phosphate Content during Adventitious Rooting

As shown in Figure 3, HRW and Glc treatments significantly increased G6P, F6P and G1P contents compared with the control. In comparison with the separate treatment of HRW and Glc, HRW + Glc treatment significantly increased the contents of G6P, F6P and G1P. However, when GlcN was added, the positive impact of HRW on the G6P, F6P and G1P contents declined significantly (Figure 3). Compared with the control, GlcN alone treatment significantly decreased the contents of G6P, F6P and G1P (Figure 3).

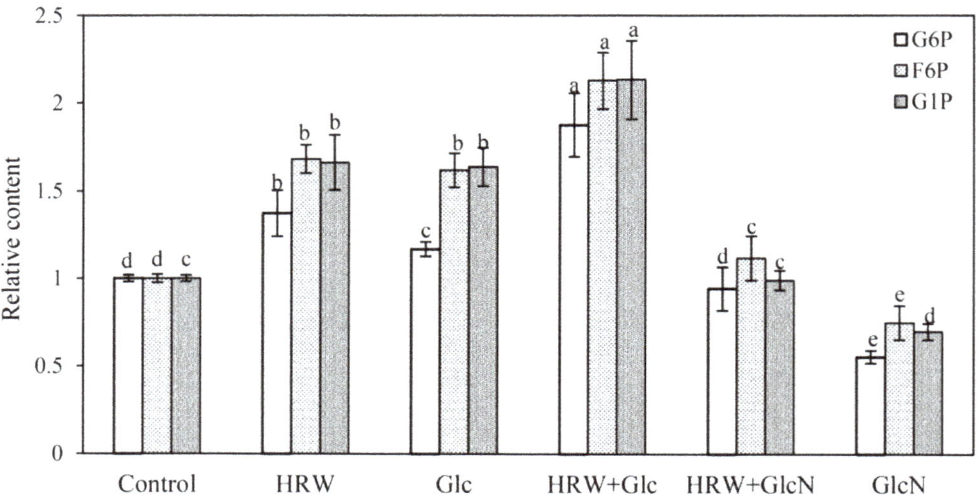

Figure 3. Changes in the contents of G6P, F6P and G1P of cucumber explants under the treatments of HRW, Glc and GlcN. n = 30 explants per replicate. Bars with different lower case letters were significantly different ($p < 0.5$). HRW: hydrogen rich water; Glc: glucose; GlcN: glucosamine; G6P: glucose-6-phosphate; F6P: fructose-6-phosphate; G1P: glucose-1-phosphate. The relative content (compared to the control) is displayed in the figure.

2.5. Effects of HRW, Glc and GlcN on Key Enzymes of Glucose Metabolism during Adventitious Rooting

Compared to the control, the activities of SS, SPS, HK, PK and AGPase were significantly enhanced by Glc or HRW (Figure 4). Meanwhile, HRW plus Glc treatment showed higher SS, SPS, HK, PK and AGPase activities than HRW or Glc treatment (Figure 4). Conversely, GlcN significantly reduced the enhancement caused by HRW. The activities of SS, SPS, HK, PK and AGPase in GlcN treatment alone were lower than that in the control (Figure 4).

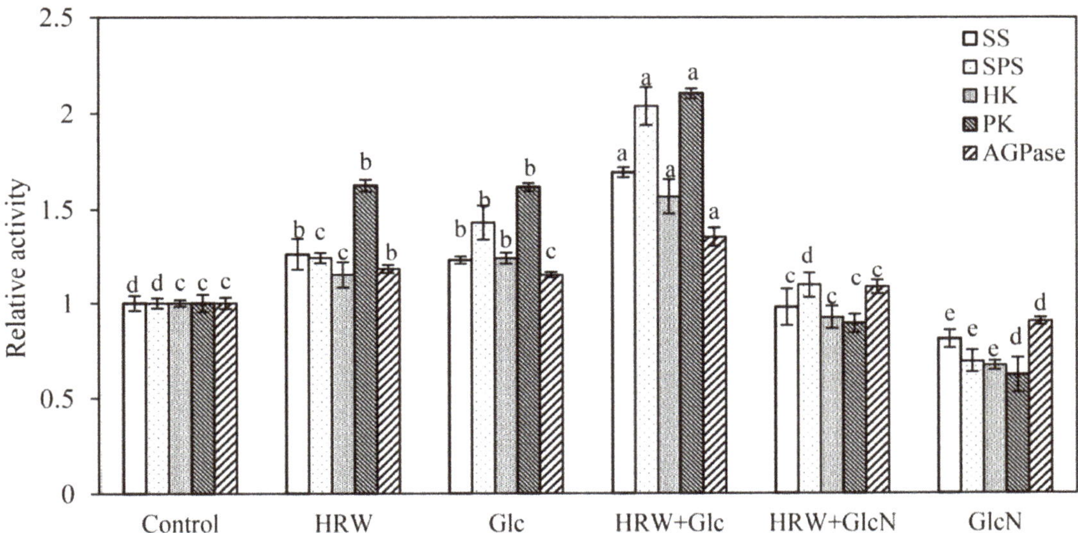

Figure 4. Effects of HRW, Glc and GlcN on SS, SPS, HK, PK and AGPase enzyme activities of cucumber explants. n = 30 explants per replicate. Bars with different lower case letters were significantly different ($p < 0.5$). HRW: hydrogen rich water; Glc: glucose; GlcN: glucosamine; SS: sucrose synthase; SPS: sucrose phosphate synthase; HK: hexokinase; PK: pyruvate kinase. AGPase: adenosine 5′-diphosphate pyrophosphorylase. The relative activity (compared to the control) is displayed in the figure.

2.6. Effects of HRW, Glc and GlcN on the Expression Levels of CsSuSy1, CsSuSy6, CsHK1, CsHK3, CsUDP1, CsUDP1-Like, CsG6P1 and CsG6P1-Like Genes during Adventitious Rooting

As shown in Figure 5, the expression levels of *CsSuSy1*, *CsSuSy6*, *CsHK1*, *CsHK3*, *CsUDP1*, *CsUDP1-like*, *CsG6P1* and *CsG6P1-like* genes in Glc or HRW treatments were significantly upregulated in comparison with the control. Additionally, the expression levels of *CsSuSy1*, *CsSuSy6*, *CsHK1*, *CsHK3*, *CsUDP1*, *CsUDP1-like*, *CsG6P1* and *CsG6P1-like* genes were the highest in HRW plus Glc treatment (Figure 5). Conversely, the expression levels of *CsSuSy1*, *CsSuSy6*, *CsHK1*, *CsHK3*, *CsUDP1*, *CsUDP1-like*, *CsG6P1* and *CsG6P1-like* genes were significantly reduced when GlcN was added (Figure 5).

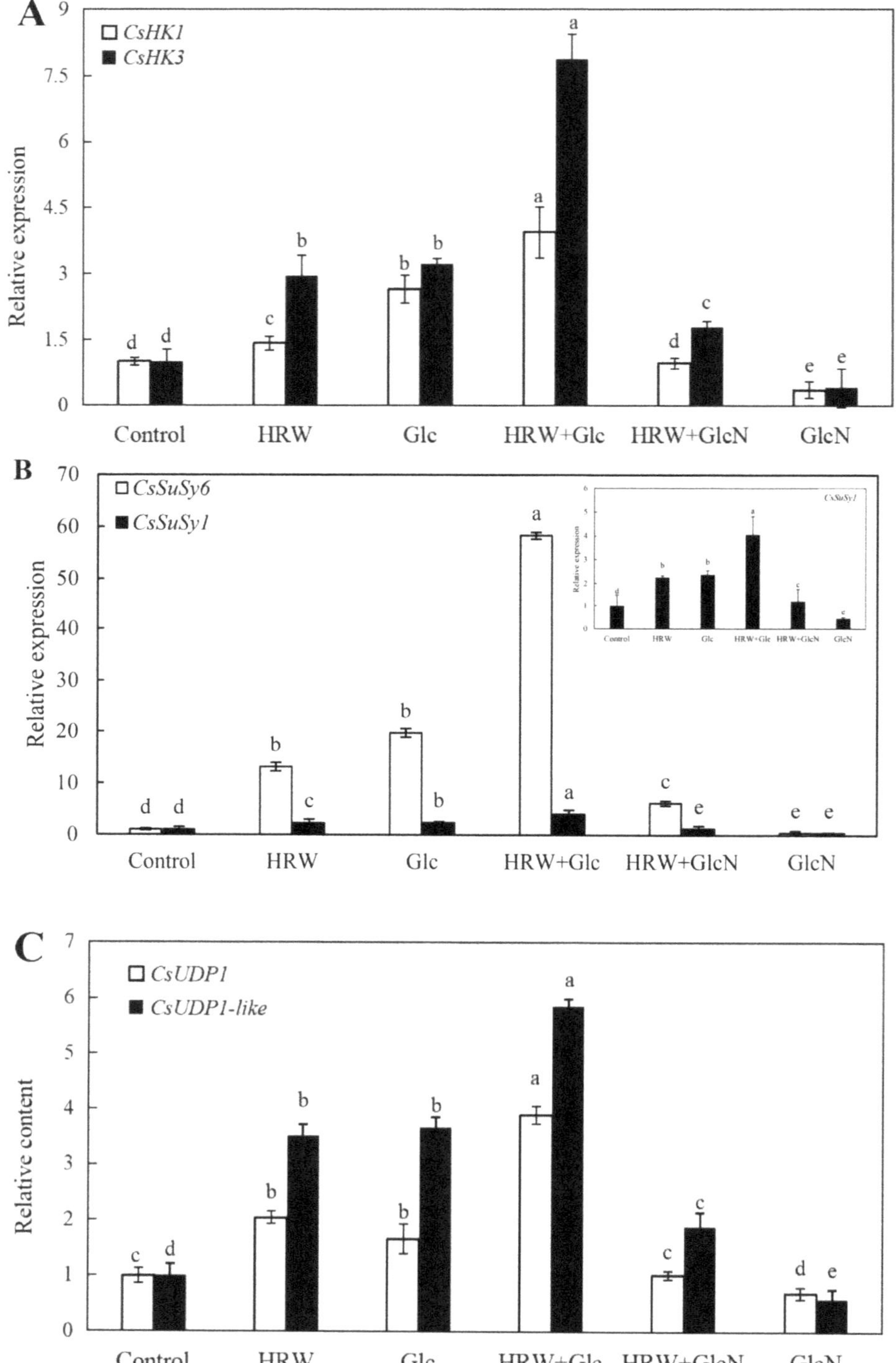

Figure 5. *Cont.*

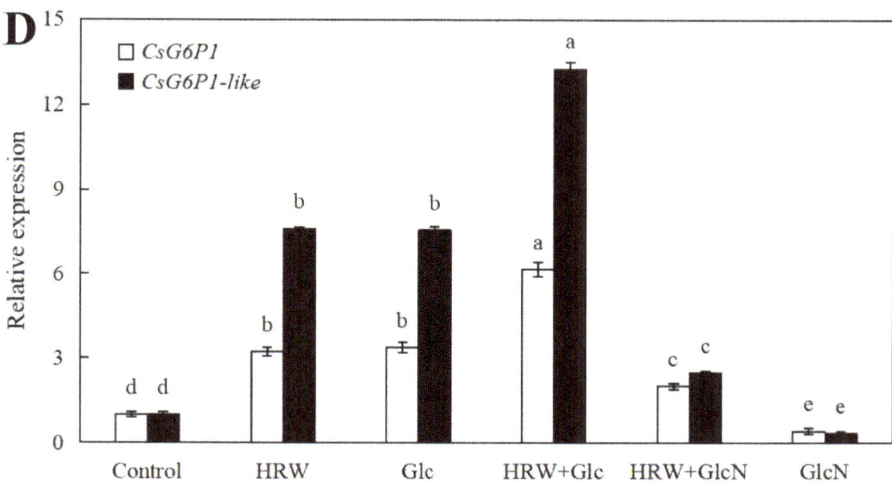

Figure 5. Effects of HRW, Glc and GlcN on the expression levels of *CsSuSy1* and *CsSuSy6* (**A**), *CsHK1* and *CsHK3* (**B**), *CsUDP1* and *CsUDP1-like* (**C**), *CsG6P1* and *CsG6P1-like* (**D**) genes in cucumber explants. n = 30 explants per replicate. Bars with different lower case letters were significantly different ($p < 0.5$). HRW: hydrogen rich water; Glc: glucose; GlcN: glucosamine.

3. Discussion

Generally, AR plays a vital role in nutrient and water absorption. Their formation is widely used for plant clonal propagation. Previous results in our lab have shown that H_2 as a positive regulator regulated adventitious rooting in cucumber [26]. However, there is little research in the crosstalk between H_2 and Glc during adventitious rooting. Here, we focus on the involvement of Glc in H_2-regulated AR formation.

Glc as signaling molecule has been the concern of by many researchers. So far, accumulating evidence indicated that Glc participated in the regulation of various growth and development, such as seed germination, seedling development and fruit ripening [15]. In the present study, we illustrated that Glc treatment promoted adventitious rooting in cucumber explants in a dose-dependent manner, and 0.10 mM Glc treatment was the most effective concentration (Table 1). Singh [27] reported that Glc controlled seedling root growth direction through regulating root waving and coiling in *Arabidopsis thaliana*, leading to altered root architecture. Mishra [24] also found that increasing Glc concentration increased root length, number of lateral roots and root hairs in *Arabidopsis thaliana* seedlings. Therefore, Glc plays a principal role in the regulation of root growth. A note of caution is that the effect of anaerobic bacteria on which glucose promoted adventitious rooting of cucumber has not been ruled out. Additionally, Zhu [16] found that NO was involved in H_2-regulated adventitious rooting of cucumber. Chen [9] reported that CO was involved in H_2-regulated adventitious rooting in cucumber under water stress. For the first time, our results indicated that Glc might be involved in H_2-regulated AR development in cucumber (Figure 1). Therefore, signal molecules including nitric oxide (NO), carbon monoxide (CO) and Glc all were required for H_2-regulated AR formation in cucumber.

In this study, we demonstrated, for the first time, that HRW treatment significantly increased the contents of glucose, sucrose, starch and total sugar during AR formation. However, the positive effects of HRW were blocked by GlcN (an inhibitor of Glu), suggesting that H_2 might regulate adventitious rooting by increasing glucose, sucrose, starch and total sugar contents (Figure 2). However, the effect of H_2 on sugar metabolism in plants has not yet been demonstrated so far. In animals, Kim and Kim [28] reported that application of HRW could improve blood glucose control for insulin deficiency and insulin resistance. HRW also has beneficial effects on lipid and glucose metabolism in humans [29].

Additionally, CH_4-medicated starch and sucrose metabolism regulated bulblet formation in *Lilium davidii* var. *unicolor* [30]. Therefore, studies on H_2 regulating glucose metabolism in plants need to be further explored. Glycolysis is a cytoplasmic pathway which breaks down glucose into G6P or F6P [17]. Glucose is trapped by phosphorylation, with the help of the enzyme hexokinase [17]. Our result found that Glc significantly increased G6P, F6P and G1P contents during adventitious rooting (Figure 3). Glc has also been shown to modulate hypocotyl directional growth in *A. thaliana* [27]. Here, we found that HRW also enhanced AR formation via increasing G6P, F6P and G1P contents. However, the positive role was inhibited by GlcN (Figure 3). Thus, H_2 regulated adventitious rooting by increasing Glu, G6P, F6P and G1P contents.

Shi [31] revealed that SPS and SS were the key enzymes in sucrose accumulation in longan fruit. HK and PK played an essential role in Glc and fructose metabolism pathway [32]. Glc, via HK-dependent and -independent signal transduction, not only regulated root growth direction of vertically grown seedlings but also caused a significant drop in bending of roots under the stimulation of gravity in *Arabidopsis Thaliana* [33]. Starch is hydrolyzed into Glc during plant metabolism. AGPase is a key enzyme governing starch synthesis. In the study, HRW or Glc significantly enhanced the SS, SPS, HK, PK and AGPase enzyme activities. Conversely, GlcN significantly inhibited the enhancement caused by HRW (Figure 4). These results suggested that H_2 promoted AR formation by enhancing the SS, SPS, HK, PK and AGPase enzyme activities. Kadowaki [34] reported that the activity of AGPase was also enhanced by sugar solution injections compared to the control, suggesting that the injection of sugar solutions is concluded to have a dual effect on root production in sweet potato. AGPase activity and production in roots was enhanced by injecting sucrose solution [35]. *SuSy1* played an important role in maintaining sucrose concentration in the cytoplasm of kiwifruit [9]. In *A. Thaliana*, AtHK1 accelerated senescence, enhanced the appearance of lateral buds and affected root growth [36]. *AtHK3* knockout mutant was found to be insensitive to 7% glucose, indicating that HK3 may play a role in sugar-sensing [37]. Transgenic rice plants overexpressing OsHXK5 or OsHXK6 exhibited growth inhibition and reduced expression of photosynthetic genes in response to glucose treatment [38]. We also found that HRW treatments were able to regulate higher expression levels of the sucrose metabolism-related genes including *CsSuSy1* and *CsSuSy6*, and glucose metabolism-related genes including *CsHK1*, *CsHK3*, *CsUDP1*, *CsUDP1-like*, *CsG6P1* and *CsG6P1-like* during AR formation (Figure 5). However, the upregulated gene expression was downregulated by GlcN (Figure 5). As mentioned above, H_2 regulated AR formation by increasing the activities of SPS, SS, HK and PK enzymes and upregulating the expression levels of *CsSuSy1*, *CsSuSy6*, *CsHK1*, *CsHK3*, *CsUDP1*, *CsUDP1-like*, *CsG6P1* and *CsG6P1-like*.

4. Materials and Methods

4.1. Plant Material

The seeds of cucumber (*Cucumis sativus* L. 'Xinchun NO. 4') were soaked in 5% sodium hypochlorite solution for disinfection and immersed in distilled water for 6 h. The seeds were then transferred to a light incubator and maintained at $25 \pm 1\ °C$ for 6 days with a 14-h photoperiod at $200\ \mu mol·s^{-1}·m^{-2}$ intensity. The 6 day-old cucumber seedlings whose primary roots were removed and used as explants were then placed in Petri dishes with distilled water or different chemicals indicated below under the same conditions of temperature and photoperiod described above for another 5 days. The treatment solutions are replaced every 8 h. Five days later, the number and length of AR per explant were measured and recorded. In addition, the explants were cultivated with different treatments under the same temperature and photoperiod conditions for 48 h. The stage belongs to the induction process without roots. Therefore, the hypocotyl base (1 cm) was collected and used for the following analysis.

4.2. The Preparation of Hydrogen-Rich Water

Purified H_2 (99.99%, v/v) was generated from a H_2-producing apparatus (QL-300, Saikesaisi Hydrogen Energy Co., Ltd., Jinan, China). Firstly, prepared hydrogen gas was bubbled into 2 L distilled water (room temperature) at a rate of 300 mL·min^{-1}. The processing was continued for 3 h. The prepared HRW was then analyzed by a dissolved hydrogen portable meter (ENH-1000, Trustlex Co., Led, Tokyo, Japan). The concentration of H_2 was 0.45 mM, which was defined as 100% hydrogen rich water (HRW). Finally, HRW was immediately diluted to the required different concentrations (0.5%, 1%, 5%, 10%, 50% and 100%).

4.3. Explant Treatments

Cucumber explants were cultivated with various concentrations of Glc (0, 0.01, 0.05, 0.10, 0.50 and 1.00 mM). In addition, GlcN as the inhibitor of glucose was used in the study; refer to Woodward [39] research. The following chemicals were carried out: control (distilled water), 50% HRW, 0.10 mM Glc, 50% HRW + 0.10 mM Glc and 0.10 µM GlcN. Each process was set to three replicates. According to the initial experimental results, the concentration of the relevant chemical substance was determined.

4.4. Determination Glucose Content

Glucose was analyzed as described by Abdellatif with some modifications [33]. Briefly, a 0.5 g sample was homogenized with 10 mL distilled water, diluted with distilled water into 50 mL volumetric flask and then mixed. The volumetric flask was then placed into a 50 °C water bath kettle for 10 min. Furthermore, 3,5-dinitrosalicylic acid (DNS) regent was prepared: 3,5-dinitrosalicylic acid (6.3 g) was dissolved in 262 mL 2 M sodium hydroxide solution. The hot-water solution of 500 mL containing 185 g potassium sodium tartrate was added. The two solutions were mixed until solvents were dissolved, followed by adding 5 g redistilled phenol and 5 g sodium sulfite, mixed and cooled to room temperature, and the volume adjusted to exactly 1000 mL with distilled water. The supernatant (2 mL) and 1.5 mL of DNS were mixed together. The mixture was centrifuged at 4000 r·min^{-1} at 4 °C for 15 min. It was then placed in a boiling water bath for 5 min, cooled to room temperature, and then filtered into a 25 mL volumetric flask. The absorbance value was measured at 540 nm and the data were recorded. The relative content (compared to the control) is shown in Figure 2.

4.5. Determination Sucrose Content

The sucrose content was analyzed according to the procedure of Tauzin [40] making appropriate modifications according to experimental requirements. The sample (1.0 g) was homogenized with 10 mL of 80% ethanol and diluted with distilled water in a 100 mL volumetric flask. The volumetric flask was placed in a water bath at a constant temperature of 80 °C for 45 min, allowed to cool to room temperature and then filtered with a quantitative analysis filter paper with a diameter of 9 cm. The filter liquor was collected as the reaction liquid. The filter (0.4 mL) was mixed with 0.2 mL 2 M NaOH and bathed for 10 min at 80 °C; after cooling to room temperature, measurement of absorbance at 540 nm was carried out. The relative content (compared to the control) is shown in Figure 2.

4.6. Determination Starch Content

The method described by Smith was adopted to determine the starch content [41]. The frozen sample (0.5 g) was ground with 2 mL distilled water. Furthermore, 3.2 mL of 60% HClO4 was added. The solution was transferred to a 10 mL tube, diluted with distilled water to volume, and mixed. The solution was then centrifuged at 5000 r·min^{-1} for 5 min, filtered and diluted with distilled water to 100 mL volumetric flask and mixed. The sucked supernatant (0.5 mL) was diluted to 3 mL with distilled water. Iodine reagent (2 mL) was added, mixed and left to stand for 5 min. Finally, it was diluted with distilled water to

10 mL. Distilled water was used as control and the absorbance was measured at 660 nm. The relative content (compared to the control) is shown in Figure 2.

4.7. Determination Total Sugar Content

Total sugar content was determined by procedures described according to Nath [42]. The 10 mL 6 M HCl was added to a 25 mL stopper tube to which 1 g sample was homogenized, and diluted to 25 mL with distilled water. The sugar was hydrolyzed during a 30 min water bath. Using phenolphthalein as its indicator, the total sugar content was determined with 6 M NaOH standard titration solution. Volume was then adjusted with distilled water to 100 mL, mixed and filtered. The filter liquor (10 mL) was transferred to a new test tube. It was then diluted to 100 mL with aqua distillate and analyzed. The absorbance value was measure at 540 nm and the data were recorded. The relative content (compared to the control) is shown in Figure 2.

4.8. Hexose Phosphate Content Measurements

Enzyme liquid was extracted by the modified method of Nägele and Wolfram [43]. In a nutshell, 0.5 g samples were ground in liquid nitrogen, then added to 2 mL 5% trichloroacetic acid (TCA) consisting of 100 mg PVPP. The mixtures were centrifuged at 12,000 r·min^{-1} for 15 min at 4 °C. The supernatant was collected, and 150 µL neutralizing buffer containing 1 M triethanolamine and 5 M potassium hydroxide (KOH) was added to the supernatant. After a reaction for 30 min on ice, the mixture was then centrifuged at 12,000 r·min^{-1} for 15 min. The supernatants were collected as the crude extract used for further analysis.

For G6P, F6P and glucose-1-phosphate (G1P), the reaction mixture consisted of 497 µL distilled water, 100 µL 1 M Hepes-KOH (pH 7.6), 100 µL 50 mM MgCl$_2$, 100 µL 4 mM NAD, 100 µL 10 mM EDTA and 100 µL of the above crude extract. The formation of blue formazan was monitored by recording the absorbance at 560 nm. G6PDH, PGI and PGM were added successively for G6P, F6P and G1P mixtures, and then were determined by measuring the absorbance at 340 nm. The relative content (compared to the control) is shown in Figure 3.

4.9. SS, SPS, HK and PK AGPase Enzymes Activity Measurement

Enzyme extracts were performed as described previously [31]. Samples were harvested, ground in liquid nitrogen and homogenized in ice-cold 9 mL PBS (pH = 7.4). Extracts were then centrifuged at 3000 r·min^{-1} for 15 min at 4 °C. The activities of sucrose synthase (SS), sucrose phosphate synthase (SPS), HK, pyruvate kinase (PK), and AGPase enzymes were measured by enzyme-linked immunosorbent assay (ELISA; AndyGene Biotechnology Co. Ltd., Beijing, China) according to the manufacturer's instructions. The relative activity (compared to the control) is shown in Figure 4

4.10. Quantitative Real-Time PCR (qRT-PCR)

Total RNA was extracted using TRIzol (Invitrogen Life Technologies) by using the method described by Huang [44]. RNA was reverse transcribed with the 5 × *Evo M-MLV* RT Master Mix (AG, China) according to the manufacturer's instructions. *Csactin* [35] was used as an internal control to calculate the relative expression. The relative transcript expression levels of the genes were quantified using $2^{-\Delta\Delta ct}$. The Cdna was amplified with 2 × SYBR Green *Pro Taq* HS Premix (AG, China) using the following primers shown in Table 2. The reactions were controlled by the following conditions: 30 s at 95 °C, then 5 s at 95 °C and 30 s at 60 °C for 40 cycles.

Table 2. Sequences of primers used for RT-PCR analysis.

Gene Symbol	Accession Number [a]	Primer Sequence (5′–3′)
CsSuSy1-F	LOC101213767	CGTGTGCTAAGGAAGGCGGAAG
CsSuSy1-R		CAGTGTCACCCCACCCTCTCTC
CsSuSy6-F	LOC101216865	TCCAACCGCCACAACTTCATCAC
CsSuSy6-R		CCATTCCCACTCTGCCCAAGC
CsHK1-F	LOC101218300	CGCCATGACCGTCGAGATGC
CsHK1-R		TTTGTACCGCCGAGATCCAATGC
CsHK3-F	LOC101215511	CACGGTCCTAGTCAGTCGGAGAG
CsHK3-R		GCCATAGCATCAACCACCTGTCTC
CsUDP1-F	LOC101206505	TCCAGAGTTCCTTGCTGAGGGTAC
CsUDP1-R		AAGCCTGAATTGCCTTGAGACCATC
CsUDP1-like-F	LOC116401645	AGTTAATGCCATTTCCGCCCTCTG
CsUDP1-like-R		TCTTGTATCCGTACCAACCGAATGC
CsG6P1-F	LOC101222586	AGGTGCGATTGCTAATCCAGATGAG
CsG6P1-R		TGCGACTTCAAGAACGAGTTAGGTG
CsG6P1-like-F	LOC101210696	AGGGTGGAGGTTTAGGGTTTAGGG
CsG6P1-like-R		GCCGCTCGTTCATTCCATTGTTC

[a] NCBI database.

4.11. Data Statistics and Analysis

Where indicated, results were expressed as the mean values $\pm$ SE of at least three independent experiments and 30 explants were taken for each replicate. Statistical analysis was performed using SPSS 22.0. For statistical analysis, Duncan's multiple test ($p < 0.05$) was chosen as appropriate.

5. Conclusions

The present study provides new insights into the roles and interactions of H_2 and Glc in the AR development in cucumber. It can be concluded that the H_2 increased the glucose, sucrose, starch, total sugar, G6P, F6P and G1P contents during adventitious rooting. Meanwhile, the sucrose-related enzymes, including SS and SPS, and glucose-related enzymes, including HK, PK and AGPase activities, were increased by H_2. Meanwhile, H_2 also upregulated the expression levels of sucrose or glucose metabolism-related genes including *CsSuSy1, CsSuSy6, CsHK1, CsHK3, CsUDP1, CsUDP1-like, CsG6P1* and *CsG6P1-like*. In conclusion, Glc might be responsible for H_2-regulated AR development.

Author Contributions: Conceptualization, W.L.; data curation, C.L. and P.H.; funding acquisition, W.L.; investigation, Z.Z.; methodology, C.L., H.L. and J.Y.; writing—original draft, Z.Z.; writing—review and editing, W.L. All authors have read and agreed to the published version of the manuscript.

Funding: This work was supported by the National Natural Science Foundation of China (Nos. 32072559, 31860568, 31560563, and 31160398); the Research Fund of Higher Education of Gansu, China (Nos. 2018C-14 and 2019B-082); and the Natural Science Foundation of Gansu Province, China (Nos. 1606RJZA073, 1606RJZA077, and 1606RJYA252).

Institutional Review Board Statement: Not applicable.

Informed Consent Statement: Not applicable.

Data Availability Statement: All data, tables and figures in this manuscript are original.

Conflicts of Interest: The authors declare no conflict of interest.

References

1. De Klerk, G.J.; Van Der Krieken, W.; De Jong, J.C. Review the formation of adventitious roots: New concepts, new possibilities. *Vitr. Cell. Dev. Biol. Plant* **1999**, *35*, 189–199. [CrossRef]
2. Li, S.W.; Leng, Y.; Feng, L.; Zeng, X.Y. Involvement of abscisic acid in regulating antioxidative defense systems and IAA-oxidase activity and improving adventitious rooting in mung bean [Vigna radiata (L.) Wilczek] seedlings under cadmium stress. *Environ. Sci. Pollut. Res.* **2014**, *21*, 525–537. [CrossRef] [PubMed]

3. Pop, T.I.; Pamfil, D.; Bellini, C. Auxin control in the formation of adventitious roots. *Not. Bot. Horti. Agrobo.* **2011**, *39*, 307–316. [CrossRef]

4. Kwak, M.S.; Kim, I.H.; Kim, S.K.; Han, T.J. Effects of brassinolide with naphthalene acetic acid on the formation of adventitious roots, trichome-like roots and calli from cultured tobacco leaf segments, and the expression patterns of CNT103. *J. Plant Biol.* **2009**, *52*, 511. [CrossRef]

5. Steffens, B.; Wang, J.X.; Sauter, M. Interactions between ethylene, gibberellin and abscisic acid regulate emergence and growth rate of adventitious roots in deepwater rice. *Planta* **2006**, *223*, 604–612. [CrossRef] [PubMed]

6. Rasmussen, A.; Hu, Y.M.; Depaepe, T.; Vandenbussche, F.; Boyer, D.F.; Geelen, D. Ethylene controls adventitious root initiation sites in arabidopsis hypocotyls independently of strigolactones. *J. Plant Growth Regul.* **2017**, *36*, 897–911. [CrossRef]

7. Takáč, T.; Obert, B.; Rolčík, J.; Šamaj, J. Improvement of adventitious root formation in flax using hydrogen peroxide. *New Biotechnol.* **2016**, *33*, 728–734. [CrossRef]

8. Jin, X.; Liao, W.B.; Yu, J.H.; Ren, P.J.; Dawuda, M.M.; Wang, M.; Niu, L.J.; Li, X.P.; Xu, X.T. Nitric oxide is involved in ethylene-induced adventitious rooting in marigold (*Tagetes erecta* L.). *CAN. J. Plant Sci.* **2017**, *97*, 620–631.

9. Chen, Y.; Wang, M.; Hu, L.L.; Liao, W.B.; Dawuda, M.M.; Li, C.L. Carbon monoxide is involved in hydrogen gas-induced adventitious root development in cucumber under simulated drought stress. *Front. Plant Sci.* **2017**, *8*, 128. [CrossRef]

10. Lin, Y.T.; Li, M.Y.; Cui, W.T.; Lu, W.; Shen, W.B. Haem oxygenase-1 is involved in hydrogen sulfide-induced cucumber adventitious root formation. *J. Plant Growth Regul.* **2012**, *31*, 519–528. [CrossRef]

11. Bai, T.H.; Dong, Z.D.; Zheng, X.B.; Song, S.W.; Jiao, J.; Wang, M.M.; Song, C.H. Auxin and its interaction with ethylene control adventitious root formation and development in apple rootstock. *Front. Plant Sci.* **2020**, *11*, 574881. [CrossRef] [PubMed]

12. Tian, R.F.; Hou, Z.; Hao, S.Y.; Wu, W.C.; Mao, X.; Tao, X.G.; Lu, T.; Liu, B.Y. Hydrogen-rich water attenuates bran damage inflammation after traumatic brain injury in rats. *Brain Res.* **2016**, *1637*, 1–13. [CrossRef] [PubMed]

13. Ostojic, S.M. Molecular hydrogen in sports medicine: New therapeutic perspectives. *Int. J. Sports Med.* **2015**, *36*, 273–279. [CrossRef]

14. Li, C.X.; Gong, T.Y.; Bian, B.T.; Liao, W.B. Roles of hydrogen gas in plants: A review. *Funct. Plant Biol.* **2018**, *45*, 783–792. [CrossRef] [PubMed]

15. Hu, H.L.; Li, P.X.; Wang, Y.N.; Gu, R.X. Hydrogen-rich water delays postharvest ripening and senescence of kiwifruit. *Food Chem.* **2014**, *156*, 100–109. [CrossRef] [PubMed]

16. Zhu, Y.C.; Liao, W.B.; Niu, L.J.; Wang, M.; Ma, Z.J. Nitric oxide is involved in hydrogen gas-induced cell cycle activation during adventitious root formation in cucumber. *BMC Plant Biol.* **2016**, *16*, 146. [CrossRef]

17. Rolland, F.; Baena-Gonzalez, E.; Sheen, J. Sugar sensing and signaling in plants: Conserved and novel mechanisms. *Annu. Rev. Plant Biol.* **2006**, *57*, 675–709. [CrossRef]

18. Xiao, W.Y.; Sheen, J.; Jang, J.C. The role of hexokinase in plant sugar signal transduction and growth and development. *Plant Mol. Biol.* **2000**, *44*, 451–461. [CrossRef]

19. Sturm, A. Invertases. Primary structures, functions, and roles in plant development and sucrose partitioning. *Plant Physiol.* **1999**, *121*, 1–8. [CrossRef]

20. Ren, X.D.; Zhang, J.J. Research progresses on the key enzymes involved in sucrose metabolism in maize. *Carbohydr. Res.* **2013**, *368*, 29–34. [CrossRef]

21. Das, A.; Rushton, P.J.; Rohila, J.S. Metabolomic profiling of soybeans (*Glycine max* L.) reveals the importance of sugar and nitrogen metabolism under drought and heat stress. *Plants* **2017**, *6*, 21. [CrossRef]

22. Gibson, S.I. Control of plant development and gene expression by sugar signaling. *Curr. Opin. Plant Biol.* **2005**, *8*, 93–102. [CrossRef]

23. Saksena, H.B.; Sharma, M.; Singh, D.; Laxmi, A. The versatile role of glucose signalling in regulating growth, development and stress responses in plants. *J. Plant Biochem. Biot.* **2020**, *29*, 687–699. [CrossRef]

24. Mishra, B.S.; Singh, M.; Aggrawal, P.; Laxmi, A. Glucose and auxin signaling interaction in controlling *Arabidopsis thaliana* seedlings root growth and development. *PLoS ONE* **2009**, *4*, e4502. [CrossRef] [PubMed]

25. Li, C.X.; Bian, B.T.; Gong, T.Y.; Liao, W.B. Comparative proteomic analysis of key proteins during abscisic acid-hydrogen peroxide-induced adventitious rooting in cucumber (Cucumis sativus L.) under drought stress. *J. Plant Physiol.* **2018**, *229*, 185–194. [CrossRef]

26. Li, C.X.; Huang, D.J.; Wang, C.L.; Wang, N.; Liao, W.B. No is involved in h$_2$-induced adventitious rooting in cucumber by regulating the expression and interaction of plasma membrane h+-atpase and 14-3-3. *Planta* **2020**, *252*, 9. [CrossRef] [PubMed]

27. Singh, M.; Gupta, A.; Laxmi, A. Glucose control of root growth direction in *Arabidopsis thaliana*. *J. Epx. Bot.* **2014**, *65*, 2981–2993. [CrossRef] [PubMed]

28. Kim, M.J.; Kim, H.K. Anti-diabetic effects of electrolyzed reduced water in streptozotocin-induced and genetic diabetic mice. *Life Sci.* **2006**, *79*, 2288–2292. [CrossRef]

29. Kajiyama, S.; Hasegawa, G.; Asano, M.; Hosoda, H.; Fukui, M.; Nakamura, N.; Kitawaki, J.; Imai, S.; Nakano, K.; Ohta, M.; et al. Supplementation of hydrogen-rich water improves lipid and glucose metabolism in patients with type 2 diabetes or impaired glucose tolerance. *Nutr. Res.* **2008**, *28*, 137–143. [CrossRef] [PubMed]

30. Qi, N.N.; Hou, X.M.; Wang, C.L.; Li, C.X.; Huang, D.J.; Li, Y.H.; Wang, N.; Liao, W.B. Methane-rich water induces bulblet formation of scale cuttings in *Lilium davidii* var. *unicolor* by regulating the signal transduction of phytohormones and their levels. *Physiol. Plant.* **2021**, *172*, 1919–1930. [CrossRef]
31. Shi, S.Y.; Wang, W.; Liu, L.Q.; Shu, B.; Wei, Y.Z.; Jue, D.W.; Fu, J.X.; Xie, J.H.; Liu, C.M. Physico-chemical properties of longan fruit during development and ripening. *Sci. Hortic.* **2016**, *207*, 160–167. [CrossRef]
32. Gupta, A.; Singh, M.; Laxmi, A. Multiple interactions between glucose and brassinosteroid signal transduction pathways in *Arabidopsis* are uncovered by Whole-Genome transcriptional profiling. *Plant Physiol.* **2015**, *168*, 1091–1105. [CrossRef]
33. Bhaji, A.; Li, J.; Ovecka, M.; Ezquer, I.; Muñoz, F.J.; Baroja-Fernández, E.; Romero, J.M.; Almagro, G.; Montero, M.; Hidalgo, M.; et al. Arabidopsis thaliana mutants lacking ADP-glucose pyrophosphorylase accumulate starch and wild-type ADP-glucose content: Further evidence for the occurrence of important sources, other than ADP-glucose pyrophosphorylase, of ADP-glucose linked to leaf starch. *Plant Cell Physiol.* **2011**, *52*, 1162–1176. [CrossRef] [PubMed]
34. Kadowaki, M.; Kubota, F.; Saitou, K. Effects of exogenous injection of different sugars on leaf photosynthesis, dry matter production and adenosine 5′-diphosphate glucose pyrophosphorylase (AGPase) activity in sweet potato, ipomoea batatas (Lam.). *J. Agron. Crop Sci.* **2010**, *186*, 37–41. [CrossRef]
35. Tsubone, M.; Kubota, F.; Saitou, K.; Kadowaki, M. Enhancement of tuberous root production and adenosine 5′-diphosphate pyrophosphorylase (AGPase) activity in sweet potato (Ipomoea batatas Lam.) by exogenous injection of sucrose solution. *J. Agron. Crop Sci.* **2000**, *184*, 181–186. [CrossRef]
36. Moore, B.; Zhou, L.; Rolland, F.; Hall, Q.; Cheng, W.H.; Liu, Y.X.; Hwang, I.; Jones, T.; Sheen, J. Role of the Arabidopsis glucose sensor HXK1 in nutrient, light, and hormonal signaling. *Science* **2003**, *300*, 332–336. [CrossRef] [PubMed]
37. Kelly, G.; David-Schwartz, R.; Sade, N.; Moshelion, M.; Levi, A.; Alchanatis, V.; Granot, D. The pitfalls of transgenic selection and new roles of AtHXK1: A high level of AtHXK1 expression uncouples hexokinase1-dependent sugar signaling from exogenous sugar. *Plant Physiol.* **2012**, *159*, 47–51. [CrossRef] [PubMed]
38. Zhang, Z.W.; Yuan, S.; Xu, F.; Yang, H.; Zhang, N.H.; Cheng, J.; Lin, H.H. The plastid hexokinase pHXK: A node of convergence for sugar and plastid signals in Arabidopsis. *FEBS Lett.* **2010**, *584*, 3573–3579. [CrossRef]
39. Woodward, G.E.; Hudson, M.T. D-glucosamine as an antagonist of glucose in carbohydrate metabolism of yeast. *J. Frankl. Inst.* **1953**, *255*, 556–560. [CrossRef]
40. Tauzin, A.S.; Giardina, T. Sucrose and invertases, a part of the plant defense, response to the biotic stresses. *Front. Plant Sci.* **2014**, *5*, 293. [CrossRef]
41. Smith, A.M.; Zeeman, S.C. Quantification of starch in plant tissues. *Nat. Protoc.* **2006**, *1*, 1342–1345. [CrossRef]
42. Nath, K.; Mahatma, M.K.; Swami, R. Role of total soluble sugar, phenols and defense related enzymes in relation to banana fruit rot by lasiodiplodia theobromae [(path.) griff.and maubl.] during ripening. *J. Plant Pathol.* **2015**, *6*, 8.
43. Nägele, T.; Weckwerth, W. Mathematical modeling reveals that metabolic feedback regulation of SnRK1 and hexokinase is sufficient to control sugar homeostasis from energy depletion to full recovery. *Plant Sci.* **2014**, *5*, 365.
44. Huang, D.J.; Li, W.T.; Dawuda, M.M.; Huo, J.Q.; Wang, C.L.; Liao, W.B. Hydrogen sulfide reduced colour change in lanzhou lily-bulb scales. *Postharvest Biol. Tec.* **2021**, *176*, 111520. [CrossRef]

Article

Molecular Hydrogen Maintains the Storage Quality of Chinese Chive through Improving Antioxidant Capacity

Ke Jiang [1,2], Yong Kuang [1], Liying Feng [1], Yuhao Liu [1], Shu Wang [1], Hongmei Du [3] and Wenbiao Shen [1,2,*]

[1] Laboratory Center of Life Sciences, College of Life Sciences, Nanjing Agricultural University, Nanjing 210095, China; 2020816131@stu.njau.edu.cn (K.J.); 2020816132@stu.njau.edu.cn (Y.K.); 10119205@njau.edu.cn (L.F.); 2019816130@njau.edu.cn (Y.L.); 2019816129@njau.edu.cn (S.W.)
[2] Center of Hydrogen Science, Shanghai Jiao Tong University, Shanghai 200240, China
[3] School of Design, Shanghai Jiao Tong University, Shanghai 200240, China; hmdu@sjtu.edu.cn
[*] Correspondence: wbshenh@njau.edu.cn; Tel.: +86-25-84-399-032; Fax: +86-25-84-396-542

Abstract: Chinese chive usually becomes decayed after a short storage time, which was closely observed with the redox imbalance. To cope with this practical problem, in this report, molecular hydrogen (H_2) was used to evaluate its influence in maintaining storage quality of Chinese chive, and the changes in antioxidant capacity were also analyzed. Chives were treated with 1%, 2%, or 3% H_2, and with air as the control, and then were stored at 4 ± 1 °C. We observed that, compared with other treatment groups, the application of 3% H_2 could significantly prolong the shelf life of Chinese chive, which was also confirmed by the obvious mitigation of decreased decay index, the loss ratio of weight, and the reduction in soluble protein content. Meanwhile, the decreasing tendency in total phenolic, flavonoid, and vitamin C contents was obviously impaired or slowed down by H_2. Results of antioxidant capacity revealed that the accumulation of reactive oxygen species (ROS) and hydrogen peroxide (H_2O_2) was differentially alleviated, which positively matched with 2,2-Diphenyl-1-picrylhydrazyl (DPPH) scavenging activity and the improved activities of antioxidant enzymes, including superoxide dismutase (SOD), guaiacol peroxidase (POD), catalase (CAT), and ascorbate peroxidase (APX). Above results clearly suggest that postharvest molecular hydrogen application might be a potential useful approach to improve the storage quality of Chinese chive, which is partially achieved through the alleviation of oxidative damage happening during the storage periods. These findings also provide potential theoretical and practical significance for transportation and consumption of perishable vegetables.

Keywords: Chinese chive; molecular hydrogen; storage quality; antioxidant capacity

Citation: Jiang, K.; Kuang, Y.; Feng, L.; Liu, Y.; Wang, S.; Du, H.; Shen, W. Molecular Hydrogen Maintains the Storage Quality of Chinese Chive through Improving Antioxidant Capacity. *Plants* **2021**, *10*, 1095. https://doi.org/10.3390/plants10061095

Academic Editor: John Hancock

Received: 26 April 2021
Accepted: 27 May 2021
Published: 29 May 2021

Publisher's Note: MDPI stays neutral with regard to jurisdictional claims in published maps and institutional affiliations.

1. Introduction

Molecular hydrogen (H_2) has emerged as a potential therapeutic medical gas in medical treatment and clinical therapy because of its selectively antioxidant capability [1]. In the last several decades, there have been several reports indicating the presence of H_2 in plants under normal or stressed conditions [2–4], although the detailed synthetic pathway(s) are not fully elucidated. Since 2012, evidence has been progressively obtained for the involvement of H_2 in plant growth and development [5,6], as well as in defense responses in plants [7,8], when challenged with salinity [9], drought [10], osmotic stress [11,12], and heavy metal exposure [13]. On the other hand, the prolonged shelf life of fruits and flowers, including tomato [14], kiwifruit [15,16], cut rose [17,18], lisianthus [19], carnation [20], by means of molecular hydrogen and magnesium hydride [21], an effective H_2-releasing material, has been discovered mainly in room temperature conditions. In most cases, the stimulation of antioxidant defense and the involvement of some other gaseous molecules, including nitric oxide [10] and hydrogen sulfide [22], as well as phytohormones [12] by exogenous H_2 are proposed as the main mechanism in plants.

Chinese chive (*Allium tuberosum* Rottler ex Spreng.) is cultivated in China and other Asian countries as well as many European countries because this popular vegetable is rich in vitamins, fiber, and sulfur compounds with antibiotic properties [23,24]. Since Chinese chive is perishable and rapidly loses freshness during transportation and storage after harvesting, this vegetable is kept at ambient temperatures or usually stored at 4 °C [25]. Previous results revealed that leaf senescence of plants, including Chinese chive, is caused by reactive oxygen species (ROS) accumulation as well as the impaired ROS scavenging system, including the reduction in superoxide dismutase (SOD), guaiacol peroxidase (POD), and catalase (CAT) activities [26–29]. Although several approaches were proposed, including the application of the cytokinin compound [29] and storage under CO_2-enriched atmospheric conditions [30], seeking more environmentally friendly and efficient methods is challenging for scientists and customers.

For the above purpose, this paper aimed to study whether or how H_2 maintains the storage quality and extends the shelf life of fresh Chinese chive, delays senescence, and affects the ROS metabolism and antioxidant defense. These obtained findings have theoretical and practical significance, and also open a new window potentially for transportation and consumption of other perishable vegetables.

2. Results

2.1. Improvement of the Visual Quality of Chive during Storage in Response to Molecular Hydrogen

Compared to freshly harvested Chinese chives, which were green and free of decay, after storing at 4 °C for 4 d, chives deteriorated rapidly, showing yellowing as well as decaying. Unlike the changes in control (4.0 ± 0.4 d) and 1% H_2 (4.0 ± 0.3 d), however, the application of 2% and 3% H_2 could prolong the shelf life of chive, with 3% H_2 showing the maximal responses (8.0 ± 0.3 d; Figure 1A).

During the time course of experiment, we also observed that visible signs of decay and wilting in chive were obviously delayed or slowed down by 2% and 3% H_2 treatments, with respect to the control samples. Comparatively, weaker responses were observed in 1% H_2-treated chive. Above results could be confirmed by the improvement of the decreased decay index and the loss ratio of weight (Figure 1B,C). For protein levels (Figure 1D), it was also observed that compared to the control group and 1% H_2 treatment, the degradation of protein in chive leaves was differentially abolished by 2% and 3% H_2 (in particular), the latter of which could be detected until 8 days of storage.

2.2. Changes of Total Phenolic and Flavonoid Contents

Figure 2 shows the changes of the total phenolic and flavonoid contents in the presence or absence of H_2 during postharvest period. For control and 1% H_2-treated group, phenolic (Figure 2A) and flavonoid (Figure 2B) contents were increased or decreased during 2 days of storage, followed by decreasing or increasing until 4 days. In the presence of 2% H_2 and 3% H_2, however, changes of total phenolic and flavonoid contents after 2 days of storage were apparently slowed down or intensified, also keeping relatively higher levels until 6 days or 8 days of storage (especially for 3% H_2 group).

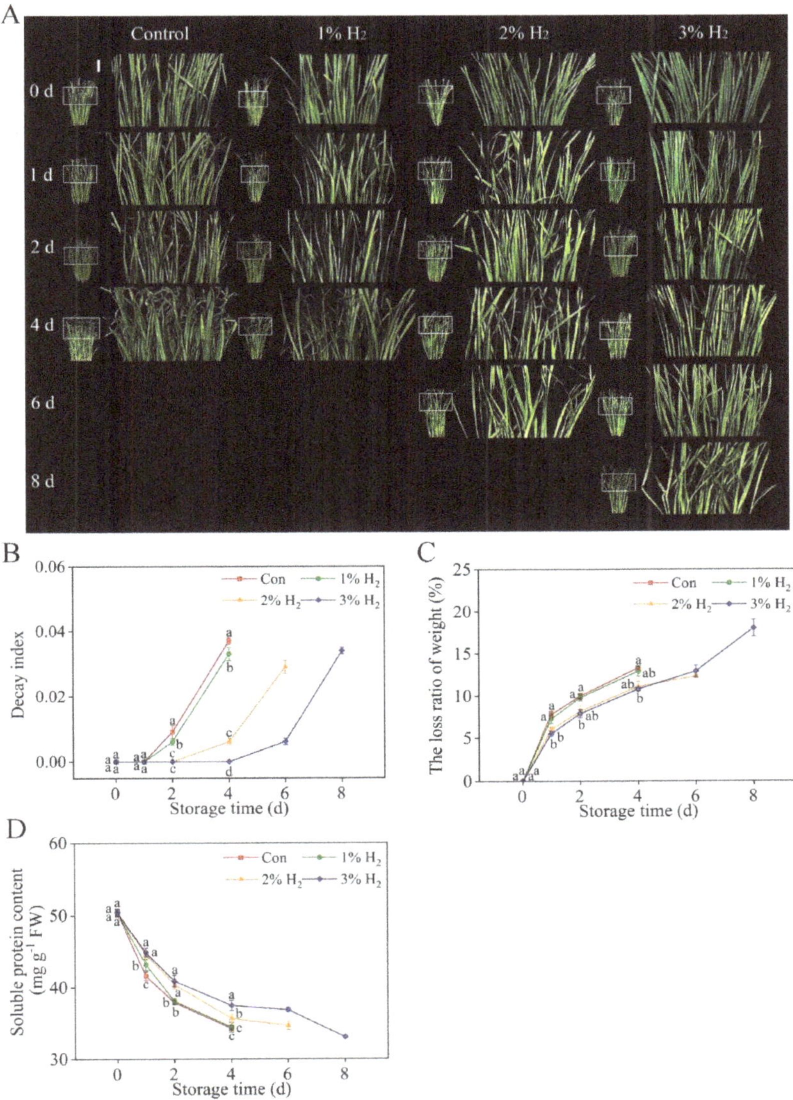

Figure 1. Effects of molecular hydrogen on shelf life of Chinese chive (**A**), decay index (**B**), the loss ratio of weight (**C**), and soluble protein content (**D**). Chinese chive was kept in air (control), 1% H_2, 2% H_2, and 3% H_2 during storage at 4 ± 1 °C. Error bars represent the standard error (SE; n = 15 for decay index, n = 15 for the loss ratio of weight, n = 5 for soluble protein content). Bars with different letters for each storage time are significantly different ($p < 0.05$) according to Duncan's multiple tests. Scale bar = 1 cm.

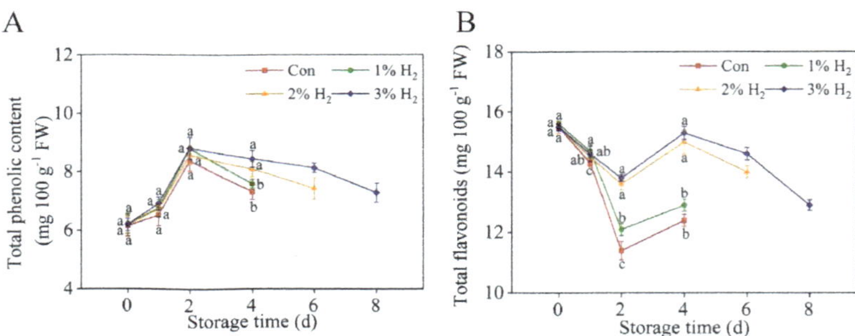

Figure 2. Time-dependent changes of total phenolic (**A**) and flavonoid (**B**) contents in response to molecular hydrogen. Chinese chive was stored at 4 ± 1 °C. Meanwhile, related parameters were analyzed. Means $\pm$ SE (n = 3 for phenolic and flavonoid contents, respectively) followed by different letters for each storage time indicate a statistical difference at $p < 0.05$.

2.3. H_2 Slowed Down the Decreased Vitamin C

A time-course analysis of vitamin C levels during chive storage was analyzed by high-performance liquid chromatography (HPLC) after treatments in the presence or absence of H_2 (Figure 3). As expected, in the control group, vitamin C contents were progressively decreased during the storage period, and were differentially abolished by 2% H_2 and 3% H_2 (in particular), lasting until 6 days or 8 days of storage. Comparatively, the application of 1% H_2 brought about a weaker but also significant change in vitamin C level at 4 days of storage. Importantly, the above rescuing effects were approximately positively matched with the biological response of H_2 in improving the visual storage quality of chive (Figure 1). Above results clearly suggest that the administration of molecular hydrogen can maintain vitamin C content in stored Chinese chive.

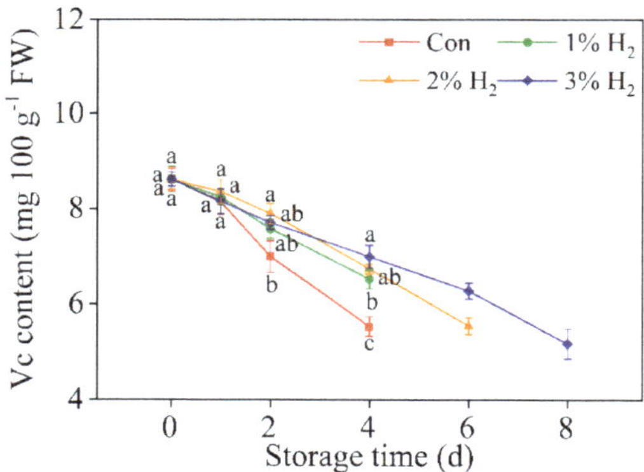

Figure 3. Time-dependent changes of vitamin C contents in response to molecular hydrogen. Chinese chive was stored at 4 ± 1 °C. Means $\pm$ SE (n = 3) followed by different letters for each storage time indicate a statistical difference at $p < 0.05$.

2.4. Redox Balance Was Reestablished by H_2

It is well known that during storage and senescence, redox imbalance occurs, which could be evaluated as the accumulation of ROS and lipid damage [26,27]. To investigate the redox status, H_2DCFDA (2′, 7′-Dichlorofluorescin diacetate), a ROS-specific fluorescent probe [10], was used to monitor ROS level in leave apex, followed by imaging by laser scanning confocal microscope (LSCM). As expected, during storage time, the fluorescence was progressively increased in the control group (until 4 days; Figure 4A,B), confirming the occurrence of redox imbalance during storage. By contrast, above fluorescence was apparently impaired or delayed until 6 days or 8 days of storage period by H_2 in a dose-dependent fashion, with 3%, in particular, reflecting the possibility that redox balance in chives might be reestablished by H_2.

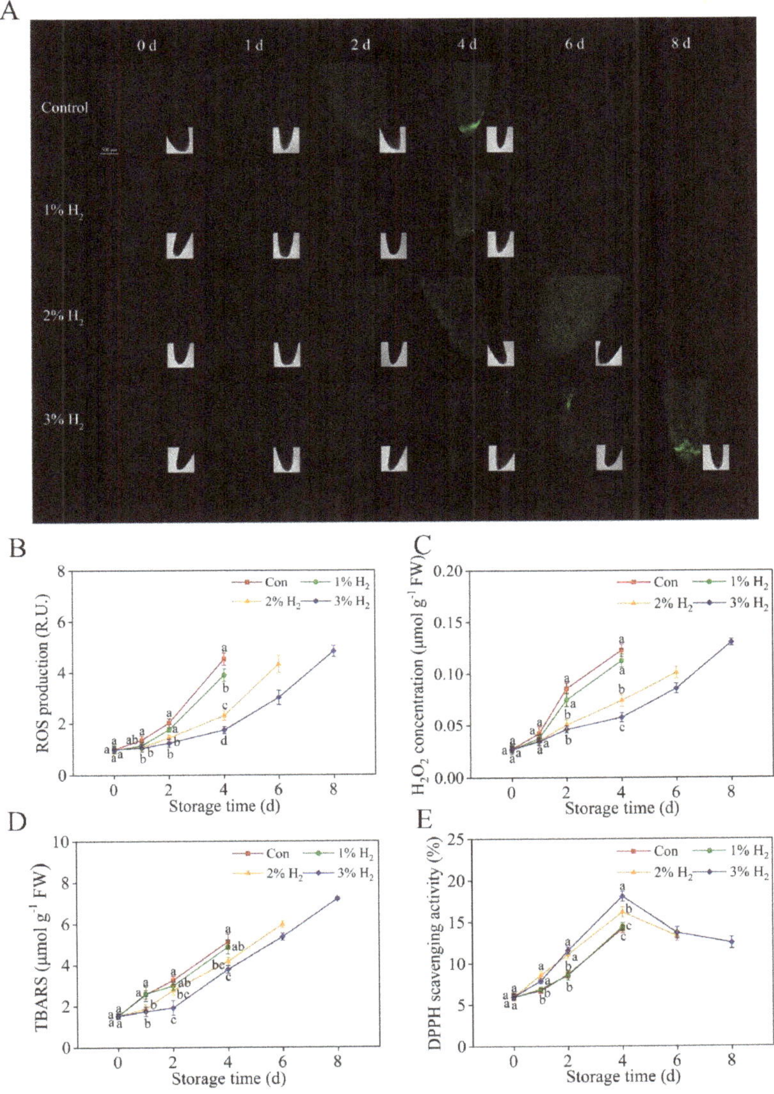

Figure 4. Redox balance was reestablished by molecular hydrogen. Chinese chive was stored at $4 \pm 1\ ^\circ C$. The LSCM images of chive leaf apex loading with H_2DCFDA (a ROS-specific fluorescent probe; (**A**)) were provided and the relative fluorescence was presented as values relative to Con at 0 day (**B**). R.U., relative units. Meanwhile, time-dependent changes in hydrogen peroxide (H_2O_2) levels (**C**), TBARS contents (**D**), and DPPH scavenging activity (**E**) of chives were determined. Means $\pm$ SE ($n = 5$ for LSCM imaging, $n = 3$ for H_2O_2 content, TBARS content, and DPPH scavenging activity, respectively) followed by different letters for each storage time indicate a statistical difference at $p < 0.05$. Scale bar = 500 μm.

To confirm the above deduction, both hydrogen peroxide (H_2O_2) and thiobarbituric acid-reactive substances (TBARS) contents (a reliable indicator of lipid damage) in leaf tissues were determined spectrophotometrically. Similar to the changes in above fluorescence, we also observed that the levels of H_2O_2 and TBARS contents during 4 days of storage period were apparently increased in a time-dependent fashion, which is partially blocked or delayed during 6 days or 8 days of storage period by different concentrations of H_2 (except 1% H_2 treatment), 3% in particular (Figure 4C,D). Consistently, results in the DPPH free radical scavenging assay demonstrated that H_2-treated groups (3% in particular) were more effective than control group, with stronger antioxidant effects during the whole storage period.

2.5. Antioxidant Enzymatic Activates Were Stimulated in the Presence of H_2

Next, to investigate whether the alleviation of oxidative damage during the storage period was causally caused by the increased antioxidant defense, activities of representative antioxidant enzymes, including SOD, POD, CAT, and ascorbate peroxidase (APX), were determined. Results shown in Figure 5 reveal that during 4 days of storage time, activities of the above four antioxidant enzymes in control sample displayed increasing tendencies (until 2 days), followed by decreasing during the rest of the storage period. Similar tendencies were observed in 1% H_2-treated group. By contrast, total activities of these enzymes were intensified until 4 days (SOD, POD, and CAT) or 2 days (APX) of storage periods, and slowed down thereafter until 8 days in other H_2 groups, and these effects were maximal in the presence of 3% H_2.

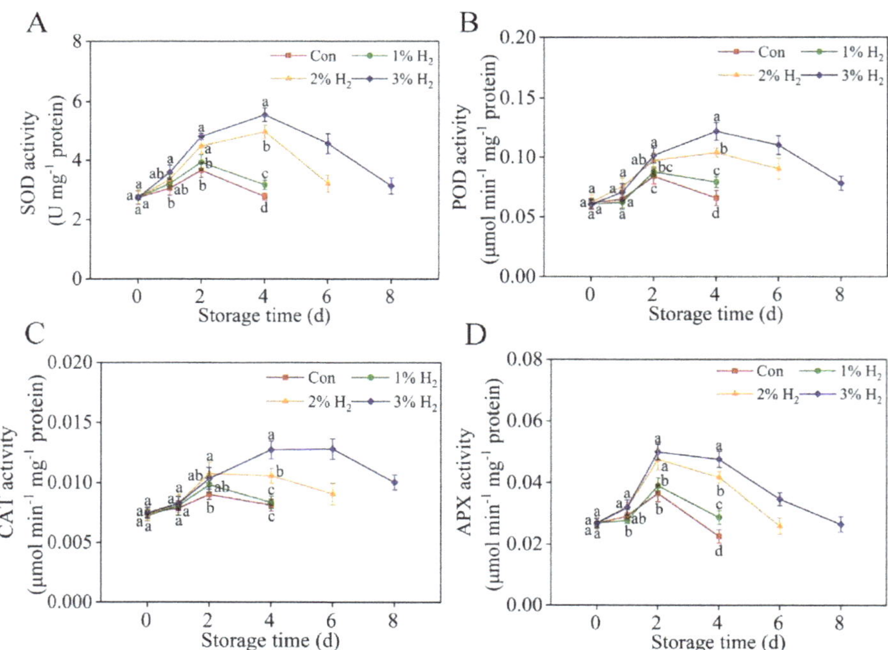

Figure 5. Time-dependent changes of SOD (**A**), POD (**B**), CAT (**C**), and APX (**D**) activities. Chinese chive was stored at 4 ± 1 °C. The enzymatic activities were expressed on a protein mass basis. Means $\pm$ SE ($n = 5$ for SOD, POD, CAT, and APX activities, respectively) followed by different letters for each storage time indicate a statistical difference at $p < 0.05$.

2.6. Changes in Reduced Glutathione (GSH) and Glutathione Reductase (GR) Activity

To determine whether the above molecular hydrogen responses resulted from influencing non-enzyme antioxidant substance and its metabolism, changes in reduced glutathione (GSH) content and glutathione reductase (GR: one of the synthetic enzymes for GSH metabolism) activity were also determined. For endogenous GSH to be tracked in situ, a commercial specific fluorescent probe monochlorobimane (MCB) was used (Figure 6A,B). As expected, in control and 1% H_2-treated group, the MCB-dependent fluorescence was increased during 2 days of storage period, followed by a decrease until 4 days. Further results illustrated that above fluorescence was obviously impaired or delayed until 6 days or 8 days of storage period by 2% and 3% (in particular) H_2 groups, reflecting the possibility that non-enzyme antioxidant substance in chives might also be influenced by H_2. Meanwhile, we observed that activities of GR displayed similar tendencies (Figure 6C).

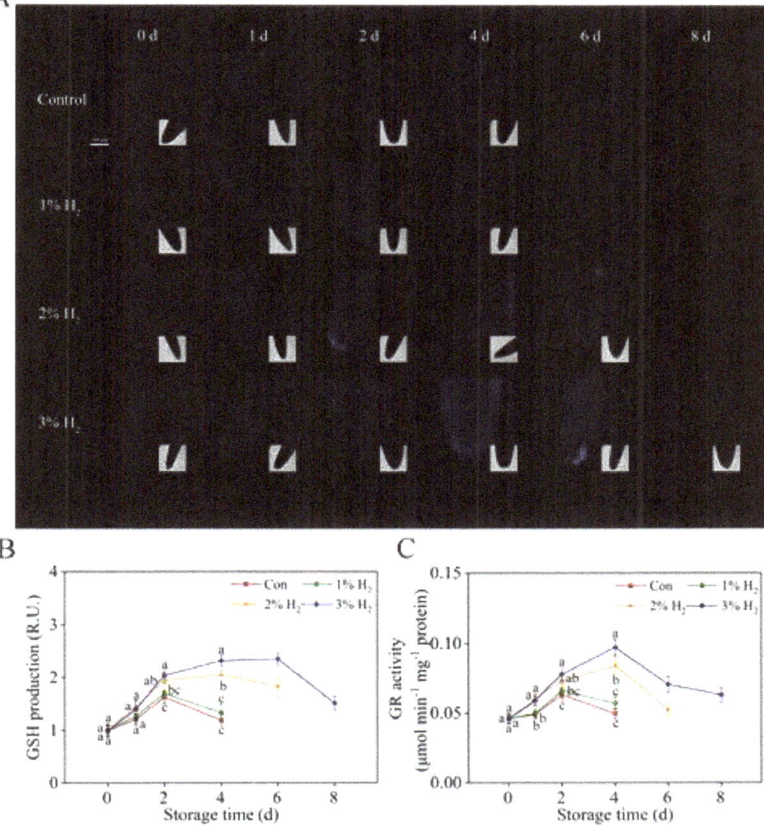

Figure 6. Changes in reduced glutathione (GSH) content and GR activity. The LSCM images of monochlorobimane (MCB)-dependent fluorescence in chive leaf apex were used to represent reduced GSH contents (**A**) and the relative fluorescence was presented as values relative to Con at 0 day (**B**). Meanwhile, time-dependent changes in GR activity (**C**) were determined. Means ± SE (n = 5 for LSCM imaging and GR activity, respectively) followed by different letters for each storage time indicate a statistical difference at p <0.05. Scale bar = 500 μm.

3. Discussion

Hydrogen-based agriculture, called hydrogen agriculture, belongs to a new low-carbon economy, which mainly refers to the application of H_2 or related storage/releasing materials for improving the production and quality of crop, forest, livestock, aquatic, and other related agricultural products during, before, and/or post-harvesting periods [5,6,8]. Here, we showed that postharvest molecular hydrogen application can maintain the storage quality of Chinese chive through improving antioxidant capacity.

Senescence is a major limiting factor in keeping chives fresh after harvesting [30]. It is well documented that both decay and weight loss are two important indexes which are closely associated with the commercial value of fruits and vegetables [31], especially for Chinese chive, which is normally only kept less than one week in open market or supermarket, even storing at 4 °C. In this report, we found that the application of 3% H_2 treatment can prolong the shelf life of chives, from 4 days of control treatment to 8 days (Figure 1A). It is a new finding. This result was further closely matched with the alleviation of the reduction in decay index (Figure 1B), the loss ratio of weight (Figure 1C), as well as the mitigation of soluble protein degradation in chives when 3% H_2 was applied (Figure 1D). Comparatively, changes in total phenolic and flavonoid contents (Figure 2) as well as vitamin C levels (Figure 3) during the storage period of chives in the presence of 3% H_2 displayed similar tendencies, and the rescuing effect in decreased vitamin C content by molecular hydrogen was previously discovered in tomato fruits after harvesting [14]. Above results clearly indicated that postharvest molecular hydrogen application could prevent the loss of nutriments during vegetable storage.

Besides vitamin C, exogenous application with hydrogen-rich water could prevent nitrite accumulation and further delay senescence of tomato fruit during storage [14]. Our study here showed similar ameliorative physiological phenotypes in Chinese chive, but for this case, we used hydrogen gas instead. Since some vegetables and fruits are prone to be perishable during storage when contacted with liquid solutions [32], our results are significant for both fundamental and applied agriculture.

Subsequently, we found that H_2 improved the preservation ability of Chinese chive by reestablishing redox homeostasis. It is well known that during storage, the occurrence of redox imbalance, caused by the accumulation of ROS, is one of the most important factors accelerating senescence during the storage of vegetables and fruits [33]. Normally, ROS accumulation is characterized by fast superoxide anion radical productive rate and therefore more H_2O_2 content, thus resulting in membrane lipid peroxidation [34,35], which is expressed as TBARS content. In our experiments, the accumulation of ROS and H_2O_2, and thereafter lipid peroxidation in leaves of chives were less pronounced in the presence of 3% H_2, especially under the similar time points, compared with the control group (Figure 4A–D), indicating that H_2 reduced the damage of excessive ROS in chives. Above results might be explained by changes in DPPH scavenging activity conferred by molecular hydrogen (Figure 4E). Redox homeostasis was therefore reestablished to some extent.

Previously, the improvement of antioxidant defense by H_2 was proposed as the main mechanism in plant response against different stresses [9,10,12]. The present work further indicated that 3% H_2 treatment significantly increased SOD activity in chives (Figure 5A). Thus, the lower production of superoxide anion radical might happen. The reduction of H_2O_2 content, one of the products of SOD, was also observed in the presence of 3% H_2 (Figure 4C). Since the scavenging of H_2O_2 is achieved by CAT, POD, and APX [13,19], activities of the above three enzymes were further analyzed. As expected, this study clearly showed that 3% H_2 treatment obviously increased their activities (Figure 5B–D), all of which further resulted in lower levels of H_2O_2 in chives (Figure 4C).

It is well known that both antioxidant enzymes and non-enzymatic antioxidant substances responsible for reestablishing redox homeostasis are responsible for delaying senescence in plants [18,20,36]. Recent results showed that GSH-related characters as well as antioxidant defense were closely related to the function of molecular hydrogen in delaying the pericarp browning of litchi [37]. Our further results discovered that, consistent with the changes in redox homeostasis (Figure 4), the administration of 3% H_2 could enhance GSH content and increase GR activity (Figure 6), one of the synthetic enzymes for GSH metabolism [38]. Importantly, these changes also matched with the beneficial phenotypes triggered by molecular hydrogen (Figure 1).

Overall, these results clearly showed that molecular hydrogen was an ideal treatment for Chinese chive storage, since 3% H_2 effectively maintained storage quality compared with the control group at 4 °C storage, which could be attributed to increased activities of antioxidant enzymes and GSH content as well as the reduced ROS accumulation in Chinese chive. Our findings therefore provide a more practical approach for transportation and consumption of some perishable vegetables and fruits.

4. Materials and Methods

4.1. Plant Materials and Treatments

Fresh Chinese chive (*Allium tuberosum* Rottler ex Spreng.) without defects, diseases and physical damage was purchased from the Suguo supermarket (Nanjing, Jiangsu Province, China; chives' place of origin: Lishui, Nanjing) and was quickly transferred to the laboratory. Afterwards, chives with uniform color and size and no tendency for withering and yellowing were selected for the subsequent experiments.

During the whole experiment, Chinese chive was stored in sealed plastic containers (1.5 L, Lock & Lock) containing air (control), and 1%, 2%, or 3% H_2. Through calculation, a certain volume of air was extracted through the injection port on each container, and then the molecular hydrogen produced by the H_2 generator was immediately injected into the container according to the corresponding volume, so as to achieve the experimental requirements. All treatment gas was renewed daily, and the plastic containers were kept in refrigerator at 4 ± 1 °C with a relative humidity (RH) of 70–75% in darkness. The sample tissue used for further analysis was the leaves of chives.

4.2. Preparation of Hydrogen Gas

Purified hydrogen gas (H_2, 99.99% (*v/v*)) was generated from H_2 generator (SHC-300; Saikesaisi Hydrogen Energy Co., Ltd., Shandong, China).

4.3. Determination of Decay Index and the Loss Ratio of Weight

According to the previous methods [29,39,40], the decay index of Chinese chive was determined based on the area percentage of the tissue affected by any decay, followed by scoring on a 1 to 5 scale; a score of 3 represented that the product was usable but not salable. For the measurement of the loss ratio of weight, the fresh weight (FW)of Chinese chive in each treatment was weighed at a fixed time every day [29,39].

4.4. Determination of Total Phenolic, Total Flavonoid

According to the previous method, total phenolic content in Chinese chive was measured by the Folin–Ciocalteu method [41] with a minor modification. First, 0.5 g of tissues was homogenized in liquid nitrogen and extracted by 80% acetone for 30 min. After centrifugation, 1 mL of the supernatant was incubated with 2 mL of undiluted Folin–Ciocalteu reagent for 2 min, and then 10% (*w/v*) Na_2CO_3 solution was added. After incubation for 1 h at 50 °C, the absorbance of the mixture was measured at 765 nm by using spectrophotometry (UV-2802 spectrophotometer, Shanghai Unico Instruments Co., Ltd., Shanghai, China). Total phenolic content was calculated from a standard curve for gallic acid, and expressed as milligrams of gallic acid per 100 g of fresh weight of sample.

Following the previous methods [41,42], the content of total flavonoid was determined by spectrophotometry. Chive tissues (0.1 g) were extracted with acetone/water/acetic acid (70:29.5:0.5, *v/v/v*) solution for 30 min. Afterwards, 4 mL of distilled water and 0.3 mL of 5% $NaNO_2$ (*w/v*) were added to 1 mL of the extract and then incubated for 5 min. Then, 0.3 mL of 10% $AlCl_3$ (*w/v*) and 2 mL of 1 mol L^{-1} NaOH were added to the mixture separately. After reaction in the dark for 15 min and centrifugation for 5 min, the absorbance was determined at 510 nm on a spectrophotometer. A standard curve was obtained by adding a variable amount of catechin. Total flavonoid content was expressed as milligrams of catechin per 100 g of FW.

4.5. Laser Scanning Confocal Microscope

According to the previous method [43], the ROS in chives were determined using a Zeiss LSM 800 confocal microscope (Carl Zeiss, Oberkochen, Germany). Briefly, the leaf apex of Chinese chive was incubated with 25 μM 2′, 7′-dichlorofluorescin diacetate (H_2DCFDA; Sigma-Aldrich, Saint Louis, America, http://www.sigmaaldrich.com (accessed on 25 April 2021)) for 20 min in the dark. After washing with HEPES/NaOH buffer (pH 7.5) three times, the sample was detected immediately by confocal microscope. Detection was performed by λ (excitation) = 488 nm and λ (emission) = 500–530 nm. The relative fluorescence was expressed as values relative to the control group at 0 day.

The reduced GSH content in chives was estimated following the previous method [44]. The leaf apex of Chinese chive was incubated with 50 μM monochlorobimane (MCB; Sigma-Aldrich, Saint Louis, America, http://www.sigmaaldrich.com (accessed on 25 April 2021)) for 20 min in the dark and was washed with HEPES buffer (pH 7.5) three times. Subsequently, the sample was detected immediately by a Zeiss LSM 800 confocal microscope (Carl Zeiss, Oberkochen, Germany; emission at 461 nm, excitation at 380 nm, respectively). The relative fluorescence was presented as values relative to Con at 0 day.

4.6. Determination of Hydrogen Peroxide (H_2O_2) Content

The H_2O_2 content was measured by spectrophotometry according to the previous method [45]. Chinese chive tissues (0.5 g) were ground with 2 mL of 0.2 mol L^{-1} $HClO_4$ on ice. After centrifugation at 12,000 $\times$ g for 15 min at 4 °C, 0.5 mL of the supernatant was mixed with 50 mmol L^{-1} H_2SO_4, 0.5 mmol L^{-1} ammonium ferrous sulfate, 200 mmol L^{-1} sorbitol, and 0.2 mmol L^{-1} xylenol orange. Afterwards, the assay reagent was incubated at 45 °C for 30 min, and the absorbance was measured at 560 nm. A standard curve was obtained by adding different amounts of H_2O_2.

4.7. Assay of Thiobarbituric Acid Reactive Substances (TBARS) Content

The content of TBARS in Chinese chive was determined by a spectrophotometer as previously described, with a minor modification [46]. The sample tissues (0.2 g) were homogenized with 2 mL of 0.1% (*w/v*) trichloroacetic acid (TCA) on ice. After centrifugation at 12,000 $\times$ g for 15 min, 0.5 mL of the supernatant was added to 1.5 mL of thiobarbituric acid (TBA). After the assay reagent was incubated at 90 °C for 20 min, the absorbance of the sample was measured at wavelengths of 532 nm, 600 nm, and 450 nm. TBARS content was expressed as μmol g^{-1} FW.

4.8. Determination of 2,2-Diphenyl-1-Picrylhydrazyl Radical (DPPH) Scavenging Activity

DPPH scavenging activity was determined according to previous method [15]. Briefly, the solution containing 10 μL of methanol extract and 3 mL of 0.1 mmol L^{-1} DPPH-methanol solution was incubated for 30 min at 25 °C in the dark. The decrease of the absorbance was measured by using a spectrophotometer at 517 nm, and blanks contained methanol instead of DPPH solution.

4.9. Determination of Vitamin C Content

The content of vitamin C in Chinese chive was estimated by using the previous methods [14,47]. The tissues of Chinese chive were derivatized with 1,2-*o*-phenylenediamine after extracting with trichloroacetic acid solution. The vitamin C content was determined by HPLC (D-2000, Hitachi, Ltd., Tokyo, Japan; excitation at 350 nm, emission at 430 nm, respectively). Quantification of vitamin C content was carried out by external calibration with *L*-ascorbic acid and expressed as mg 100 g^{-1} FW.

4.10. Assay of Antioxidant Enzyme Activity

In order to determine the activities of SOD, POD, CAT, APX, and GR, fresh tissue samples (0.2 g) from Chinese chive were homogenized in 3 mL of 50 mmol L^{-1} phosphate buffer (pH 7.0) [19]. SOD activity was determined by detecting the inhibition of photochemical reduction of nitro blue tetrazolium (NBT) at 560 nm according to the previous methods [19,48], and one enzyme unit (U) was considered to be the amount of enzyme corresponding to 50% inhibition of NBT reduction. POD activity was determined by following the oxidation of guaiacol at 470 nm [13]. CAT activity was estimated by detecting the reduction of H_2O_2 at 240 nm [48]. APX activity was assayed by monitoring the decrease at 290 nm after adding 1 mmol L^{-1} ascorbic acid [13]. GR activity was determined by monitoring the oxidation of nicotinamide adenine dinucleotide phosphate at 340 nm [49]. The enzyme activity was expressed on a protein mass basis, and the protein concentration of Chinese chive was determined by BCA (bicinchoninic acid) Protein Assay Kit (TaKaRa Bio Inc., Dalian, China).

4.11. Experimental Design

Following the previous reports [19,50] with some minor modifications, all experiments were arranged in a randomized complete block design. The experiment was carried out 3 times with triplicates per experiment, and each replicate included 15 Chinese chives. In order to determine decay index and the loss ratio of weight, 15 chives were selected for determination each time, and the total number of three repeated chives was 45 (15 × 3), and the representative phenotypes were photographed. Five chives per replicate were selected for the evaluation of LSCM imaging, soluble protein content, and enzyme activities, and the total of three repeated chives was 15 (5 × 3). For other parameters, including total phenolic and flavonoid contents, vitamin C content, H_2O_2 concentration, TBARS content, and DPPH scavenging activity, three chives were selected for each repetition, and the total of the three repetitions was 9 (3 × 3). All samples used for the determination were taken from healthy tissues of the samples.

4.12. Statistical Analysis

All values in this study were expressed as the means ± standard error (SE) from three independent experiments with three biological replicates for each. Data analysis was performed by using SPSS 16.0 software (SPSS Inc., Chicago, IL, USA), and one-way analysis of variance (ANOVA) was used to analyze differences among treatments according to Duncan's multiple range test, and $p < 0.05$ as significant.

Author Contributions: K.J. and Y.K. conceived and designed the experiments. K.J., Y.K., L.F., Y.L., and S.W. performed the research. K.J., L.F., Y.L., and S.W. analyzed the data. K.J., L.F., H.D., and W.S. wrote the paper. All authors have read and agreed to the published version of the manuscript.

Funding: This work was supported by the Foshan Agriculture Science and Technology Project (Foshan City Budget, No. 140, 2019) and the Funding from Center of Hydrogen Science, Shanghai Jiao Tong University, China.

Institutional Review Board Statement: Not applicable.

Informed Consent Statement: Not applicable.

Data Availability Statement: All data reported here is available from the authors upon request.

Conflicts of Interest: The authors declare that they have no competing interests.

Abbreviations

Abbreviations	
APX	ascorbate peroxidase
ANOVA	one-way analysis of variance
BCA	bicinchoninic acid
CAT	catalase
Con	control
DPPH	2,2-Diphenyl-1-picrylhydrazyl
FW	fresh weight
GSH	glutathione
GR	glutathione reductase
H_2	molecular hydrogen
H_2DCFDA	$2',7'$-Dichlorofluorescin diacetate
HPLC	high-performance liquid chromatography
H_2O_2	hydrogen peroxide
LSCM	laser scanning confocal microscope
MCB	monochlorobimane
NBT	nitro blue tetrazolium
POD	guaiacol peroxidase
RH	relative humidity
ROS	reactive oxygen species
R.U.	relative units
SE	standard error
SOD	superoxide dismutase
TBARS	thiobarbituric acid reactive substances

References

1. Ohsawa, I.; Ishikawa, M.; Takahashi, K.; Watanabe, M.; Nishimaki, K.; Yamagata, K.; Katsura, K.; Katayama, Y.; Asoh, S ; Ohta, S. Hydrogen acts as a therapeutic antioxidant by selectively reducing cytotoxic oxygen radicals. *Nat. Med.* **2007**, *13*, 688–694. [CrossRef]
2. Boichenko, E.A. Hydrogenase from isolated chloroplasts. *Biokhimiya* **1947**, *12*, 153–162.
3. Renwick, G.M.; Giumarro, C.; Siegel, S.M. Hydrogen metabolism in higher plants. *Plant Physiol.* **1964**, *39*, 303–306. [CrossRef] [PubMed]
4. Zeng, J.Q.; Zhang, M.Y.; Sun, X.J. Molecular hydrogen is involved in phytohormone signaling and stress responses in plants. *PLoS ONE* **2013**, *8*, e71038. [CrossRef] [PubMed]
5. Shen, W.B.; Sun, X.J. Hydrogen biology: It is just beginning. *Chin. J. Biochem. Mol. Biol.* **2019**, *35*, 1037–1050.
6. Wang, Y.Q.; Liu, Y.H.; Wang, S.; Du, H.M.; Shen, W.B. Hydrogen agronomy: Research progress and prospects. *J. Zhejiang Univ. Sci. B* **2020**, *21*, 841–855. [CrossRef] [PubMed]
7. Russell, G.; Zulficiar, F.; Hancock, J.T. Hydrogenases and the role of molecular hydrogen in plants. *Plants* **2020**, *9*, 1136. [CrossRef]
8. Li, L.N.; Lou, W.; Kong, L.S.; Shen, W.B. Hydrogen commonly applicable from medicine to agriculture: From molecular mechanisms to the field. *Curr. Pharm. Des.* **2021**, *27*, 747–759. [CrossRef]
9. Xie, Y.J.; Mao, Y.; Lai, D.W.; Zhang, W.; Shen, W.B. H_2 Enhances Arabidopsis salt tolerance by manipulating ZAT10/12-mediated antioxidant defence and controlling sodium exclusion. *PLoS ONE* **2012**, *7*, e49800. [CrossRef]
10. Xie, Y.J.; Mao, Y.; Zhang, W.; Lai, D.W.; Wang, Q.Y.; Shen, W.B. Reactive oxygen species-dependent nitric oxide production contributes to hydrogen-promoted stomatal closure in Arabidopsis. *Plant Physiol.* **2014**, *165*, 759–773. [CrossRef]
11. Jin, Q.J.; Cui, W.T.; Dai, C.; Zhu, K.K.; Zhang, J.; Wang, R.; La, H.G.; Li, X.; Shen, W.B. Involvement of hydrogen peroxide and heme oxygenase-1 in hydrogen gas-induced osmotic stress tolerance in alfalfa. *Plant Growth Regul.* **2016**, *80*, 215–223. [CrossRef]
12. Felix, K.; Su, J.C.; Lu, R.F.; Zhao, G.; Cui, W.T.; Wang, R.; Mu, H.L.; Cui, J.; Shen, W.B. Hydrogen-induced tolerance against osmotic stress in alfalfa seedlings involves ABA signaling. *Plant Soil* **2019**, *445*, 409–423. [CrossRef]
13. Cui, W.T.; Gao, C.Y.; Fang, P.; Lin, G.Q.; Shen, W.B. Alleviation of cadmium toxicity in Medicago sativa by hydrogen-rich water. *J. Hazard. Mater.* **2013**, *260*, 715–724. [CrossRef]
14. Zhang, Y.H.; Zhao, G.; Cheng, P.F.; Yan, X.Y.; Li, Y.; Cheng, D.; Wang, R.; Chen, J.; Shen, W.B. Nitrite accumulation during storage of tomato fruit as prevented by hydrogen gas. *Int. J. Food Prop.* **2019**, *22*, 1425–1438. [CrossRef]
15. Hu, H.L.; Li, P.X.; Wang, Y.N.; Gu, R.X. Hydrogen-rich water delays postharvest ripening and senescence of kiwifruit. *Food Chem.* **2014**, *156*, 100–109. [CrossRef] [PubMed]
16. Hu, H.L.; Zhao, S.P.; Li, P.X.; Shen, W.B. Hydrogen gas prolongs the shelf life of kiwifruit by decreasing ethylene biosynthesis. *Postharvest Biol. Technol.* **2018**, *135*, 123–130. [CrossRef]

17. Wang, C.L.; Fang, H.; Gong, T.Y.; Zhang, J.; Niu, L.J.; Huang, D.J.; Huo, J.Q.; Liao, W.B. Hydrogen gas alleviates postharvest senescence of cut rose 'Movie star' by antagonizing ethylene. *Plant Mol. Biol.* **2020**, *102*, 271–285. [CrossRef]

18. Li, Y.; Li, L.N.; Wang, S.; L, Y.H.; Zou, J.X.; Ding, W.J.; Du, H.M.; Shen, W.B. Magnesium hydride acts as a convenient hydrogen supply to prolong the vase life of cut roses by modulating nitric oxide synthesis. *Postharvest Biol. Technol.* **2021**, *177*, 111526. [CrossRef]

19. Su, J.C.; Nie, Y.; Zhao, G.; Cheng, D.; Wang, R.; Chen, J.; Zhang, S.H.; Shen, W.B. Endogenous hydrogen gas delays petal senescence and extends the vase life of lisianthus cut flowers. *Postharvest Biol. Technol.* **2019**, *147*, 148–155. [CrossRef]

20. Li, L.N.; Liu, Y.H.; Wang, S.; Zou, J.X.; Ding, W.J.; Shen, W.B. Magnesium hydride-mediated sustainable hydrogen supply prolongs the vase life of cut carnation flowers via hydrogen sulfide. *Front. Plant Sci.* **2020**, *11*, 595376. [CrossRef] [PubMed]

21. Xia, G.L.; Zhang, L.J.; Chen, X.W.; Huang, Y.Q.; Sun, D.L.; Fang, F.; Guo, Z.P.; Yu, X.B. Carbon hollow nanobubbles on porous carbon nanofibers: An ideal host for high-performance sodium-sulfur batteries and hydrogen storage. *Energy Storage Mater.* **2018**, *14*, 314–323. [CrossRef]

22. Zhang, Y.H.; Cheng, P.F.; Wang, Y.Q.; Li, Y.; Su, J.C.; Chen, Z.P.; Yu, X.L.; Shen, W.B. Genetic elucidation of hydrogen signaling in plant osmotic tolerance and stomatal closure via hydrogen sulfide. *Free. Radic. Biol. Med.* **2020**, *161*, 1–14. [CrossRef]

23. Imahori, Y.; Suzuki, Y.; Uemura, K.; Kishioka, I.; Fujiwara, H.; Ueda, Y.; Chachin, K. Physiological and quality responses of Chinese chive leaves to low oxygen atmosphere. *Postharvest Biol. Technol.* **2004**, *31*, 295–303. [CrossRef]

24. Hu, G.H.; Lu, Y.H.; Wei, D.Z. Chemical characterization of Chinese chive seed (*Allium tuberosum* Rottl.). *Food Chem.* **2005**, *99*, 693–697. [CrossRef]

25. Ishii, K.; Okubo, M. The keeping quality of Chinese chive (*Allium tuberosum* Rottler) by low temperature and seal-packaging with polyethylene bag. *J. Jpn. Soc. Hortic. Sci.* **1984**, *53*, 87–95. [CrossRef]

26. Borrel, A.; Carbonell, L.; Farràs, R.; PuigParcllada, P.; Tiburcio, A.F. Polyamines inhibit lipid peroxidation in senescing oat leaves. *Physiol. Plant.* **1997**, *99*, 385–390. [CrossRef]

27. Wen-biao, S.; Mao-bing, Y.; Lang-lai, X.; Rong-xian, Z. Changes of ability of scavenging active oxygen during natural senescence of wheat flag leaves. *J. Integr. Plant Biol.* **1997**, *39*, 634–640.

28. Rogers, H.; Munné-Bosch, S. Production and scavenging of reactive oxygen species and redox signaling during leaf and flower senescence: Similar but different. *Plant Physiol.* **2016**, *171*, 1560–1568. [CrossRef] [PubMed]

29. Jia, L.E.; Liu, S.; Duan, X.M.; Zhang, C.; Wu, Z.H.; Liu, M.C.; Guo, S.G.; Zuo, J.H.; Wang, L.B. 6-Benzylaminopurine treatment maintains the quality of Chinese chive (*Allium tuberosum* Rottler ex Spreng.) by enhancing antioxidant enzyme activity. *J. Integr. Agric.* **2017**, *16*, 1968–1977. [CrossRef]

30. Imahori, Y.; Suzuki, Y.; Kawagishi, M.; Ishimaru, M.; Ueda, Y.; Chachin, K. Physiological responses and quality attributes of Chinese chive leaves exposed to CO_2-enriched atmospheres. *Postharvest Biol. Technol.* **2007**, *46*, 160–166. [CrossRef]

31. Tzortzakis, N.; Chrysargyris, A. Postharvest ozone application for the preservation of fruits and vegetables. *Food Rev. Int.* **2017**, *33*, 270–315. [CrossRef]

32. Jiang, Y.M.; Zhu, X.R.; Li, Y.B. Postharvest control of litchi fruit rot by *Bacillus subtilis. LWT-Food Sci. Technol.* **2001**, *34*, 430–436. [CrossRef]

33. Wu, Z.F.; Tu, M.M.; Yang, X.P.; Xu, J.H.; Yu, Z.F. Effect of cutting on the reactive oxygen species accumulation and energy change in postharvest melon fruit during storage. *Sci. Hortic.* **2019**, *257*, 108752. [CrossRef]

34. Gao, H.J.; Zhao, F.; Yang, J.X.; Yang, H.Q. Nitric oxide alleviates lipid peroxidation induced by osmotic stress during senescence of detached leaves of Malus hupehensis Rehd. *J. Hortic. Sci. Biotechnol.* **2010**, *85*, 367–373. [CrossRef]

35. Imahori, Y.; Bai, J.H.; Baldwin, E. Antioxidative responses of ripe tomato fruit to postharvest chilling and heating treatments. *Sci. Hortic.* **2016**, *198*, 398–406. [CrossRef]

36. Wang, P.; Sun, X.; Li, C.; Wei, Z.W.; Liang, D.; Ma, F.W. Long-term exogenous application of melatonin delays drought-induced leaf senescence in apple. *J. Pineal Res.* **2013**, *54*, 292–302. [CrossRef]

37. Yun, Z.; Gao, H.J.; Chen, X.; Chen, Z.S.Z.; Zhang, Z.K.; Li, T.T.; Qu, H.X.; Jiang, Y.M. Effects of hydrogen water treatment on antioxidant system of litchi fruit during the pericarp browning. *Food Chem.* **2021**, *336*, 127618. [CrossRef]

38. Couto, N.; Wood, J.; Barber, J. The role of glutathione reductase and related enzymes on cellular redox homoeostasis network. *Free Radic. Biol. Med.* **2016**, *95*, 27–42. [CrossRef] [PubMed]

39. Siddiqui, M.W.; Dutta, P.; Dhua, R.S.; Dey, A. Changes in biochemical composition of mango in response to pre-harvest gibberellic acid spray. *Agric. Conspec. Sci.* **2014**, *78*, 331–335.

40. Cantwell, M.I.; Thangaiah, A. Acceptable cooling delays for selected warm season vegetables and melons. In *XXVIII International Horticultural Congress on Science and Horticulture for People (IHC2010): International Symposium on Postharvest Technology in the Global Market*; Cantwell, M.I., Almeida, D.P.F., Eds.; International Society for Horticultural Science: Leuven, Belgium, 2012; Volume 2, pp. 77–34.

41. Zhang, X.Y.; Wei, J.Y.; Tian, J.Y.; Li, N.; Jia, L.; Shen, W.B.; Cui, J. Enhanced anthocyanin accumulation of immature radish microgreens by hydrogen-rich water under short wavelength light. *Sci. Hortic.* **2019**, *247*, 75–85. [CrossRef]

42. Hasperué, J.H.; Rodoni, L.M.; Guardianelli, L.M.; Chaves, A.R.; Martinez, G.A. Use of LED light for Brussels sprouts postharvest conservation. *Sci. Hortic.* **2016**, *213*, 281–286. [CrossRef]

43. Chen, Z.P.; Xie, Y.J.; Gu, Q.; Zhao, G.; Zhang, Y.H.; Cui, W.T.; Xu, S.; Wang, R.; Shen, W.B. The *AtrbohF*-dependent regulation of ROS signaling is required for melatonin-induced salinity tolerance in *Arabidopsis*. *Free Radic. Biol. Med.* **2017**, *108*, 465–477. [CrossRef]
44. Gu, Q.; Chen, Z.P.; Cui, W.T.; Zhang, Y.H.; Hu, H.L.; Yu, X.L.; Wang, Q.Y.; Shen, W.B. Methane alleviates alfalfa cadmium toxicity via decreasing cadmium accumulation and reestablishing glutathione homeostasis. *Ecotoxicol. Environ. Saf.* **2018**, *147*, 861–871. [CrossRef]
45. Gu, Q.; Chen, Z.P.; Yu, X.L.; Cui, W.T.; Pan, J.C.; Zhao, G.; Xu, S.; Wang, R.; Shen, W.B. Melatonin confers plant tolerance against cadmium stress via the decrease of cadmium accumulation and reestablishment of microRNA-mediated redox homeostasis. *Plant Sci.* **2017**, *261*, 28–37. [CrossRef]
46. Hodges, D.M.; DeLong, J.M.; Forney, C.F.; Prange, R.K. Improving the thiobarbituric acid-reactive-substances assay for estimating lipid peroxidation in plant tissues containing anthocyanin and other interfering compounds. *Planta* **1999**, *207*, 604–611. [CrossRef]
47. Georgé, S.; Tourniaire, F.; Gautier, H.; Goupy, P.; Rock, E.; Caris-Veyrat, C. Changes in the contents of carotenoids, phenolic compounds and vitamin C during technical processing and lyophilisation of red and yellow tomatoes. *Food Chem.* **2011**, *124*, 1603–1611. [CrossRef]
48. Su, J.C.; Zhang, Y.H.; Nie, Y.; Cheng, D.; Wang, R.; Hu, H.L.; Chen, J.; Zhang, J.F.; Du, Y.W.; Shen, W.B. Hydrogen-induced osmotic tolerance is associated with nitric oxide-mediated proline accumulation and reestablishment of redox balance in alfalfa seedlings. *Environ. Exp. Bot.* **2018**, *147*, 249–260. [CrossRef]
49. Schaedle, M. Chloroplast glutathione reductase. *Plant Physiol.* **1977**, *59*, 1011–1012. [CrossRef]
50. In, B.C.; Ha, S.T.T.; Lee, Y.S.; Lim, J.H. Relationships between the longevity, water relations, ethylene sensitivity, and gene expression of cut roses. *Postharvest Biol. Technol.* **2017**, *131*, 74–83. [CrossRef]

Article

Degradation of Carbendazim by Molecular Hydrogen on Leaf Models

Tong Zhang [1], Yueqiao Wang [1], Zhushan Zhao [1], Sheng Xu [2] and Wenbiao Shen [1,*]

[1] Laboratory Center of Life Sciences, College of Life Sciences, Nanjing Agricultural University, Nanjing 210095, China; 2020116098@stu.njau.edu.cn (T.Z.); 2019116098@njau.edu.cn (Y.W.); 2021816131@stu.njau.edu.cn (Z.Z.)

[2] Institute of Botany, Jiangsu Province and Chinese Academy of Sciences, Nanjing 210014, China; xusheng@cnbg.net

* Correspondence: wbshenh@njau.edu.cn; Tel.: +86-25-84-399-032; Fax: +86-25-84-396-542

Abstract: Although molecular hydrogen can alleviate herbicide paraquat and *Fusarium* mycotoxins toxicity in plants and animals, whether or how molecular hydrogen influences pesticide residues in plants is not clear. Here, pot experiments in greenhouse revealed that degradation of carbendazim (a benzimidazole pesticide) in leaves could be positively stimulated by molecular hydrogen, either exogenously applied or with genetic manipulation. Pharmacological and genetic increased hydrogen gas could increase glutathione metabolism and thereafter carbendazim degradation, both of which were abolished by the removal of endogenous glutathione with its synthetic inhibitor, in both tomato and in transgenic *Arabidopsis* when overexpressing the hydrogenase 1 gene from *Chlamydomonas reinhardtii*. Importantly, the antifungal effect of carbendazim in tomato plants was not obviously altered regardless of molecular hydrogen addition. The contribution of glutathione-related detoxification mechanism achieved by molecular hydrogen was confirmed. Our results might not only illustrate a previously undescribed function of molecular hydrogen in plants, but also provide an environmental-friendly approach for the effective elimination or reduction of pesticides residues in crops when grown in pesticides-overused environmental conditions.

Keywords: hydrogen gas; carbendazim degradation; glutathione metabolism; detoxification system; redox balance

Citation: Zhang, T.; Wang, Y.; Zhao, Z.; Xu, S.; Shen, W. Degradation of Carbendazim by Molecular Hydrogen on Leaf Models. *Plants* **2022**, *11*, 621. https://doi.org/10.3390/plants11050621

Academic Editor: John Hancock

Received: 7 February 2022
Accepted: 23 February 2022
Published: 25 February 2022

Publisher's Note: MDPI stays neutral with regard to jurisdictional claims in published maps and institutional affiliations.

1. Introduction

Pesticides are indispensable for sustained food production [1]. However, excessive use of pesticides, including fungicides, could result in pesticide pollution of vegetables and environmental contamination. More importantly, fungicides can enter the human body and the food chain through foods, polluted air, and water. Therefore, human health is currently being threatened by the continued use of fungicides [2]. Carbendazim (CAR; methyl 1H-benzimidazol-2-ylcarbamate), a benzimidazole fungicide, is generally used in treatment and control of fungal diseases in vegetables, fruits, flowers, etc. [3]. Since β-tubulin is the target of CAR, its antifungal mechanism could be contributed to interfere with the formation of the spindle during the mitosis of pathogenic bacteria [4].

Normally, CAR remains on the plant surface, or is absorbed by plants and then accumulates at the end of the food chain, ultimately posing a serious threat to human health. Importantly, excess CAR has been reported to disrupt the human endocrine system. At low concentrations, this chemical can even damage the mammalian liver, reproductive tissues, and endocrine [5]. Therefore, there is an urgent need to develop effective strategies to reduce CAR residues in agricultural products.

The degradation of CAR is mainly photochemical catalytic degradation and biodegradation, but the process is relatively slow [6]. Thus, research on accelerating the degradation of carbendazim in plants and the environment has attracted more attention, especially with an environmental-friendly approach. For example, previous research discovered the nitric oxide (NO), a natural by-product of nitrogen metabolism [7], can participate in the degradation process of chlorothalonil, a broad-spectrum chlorine fungicide, achieved by brassinosteroid [8].

Hydrogen gas (H_2) is extremely small and was previously regarded as a relatively inert molecule. During the last decade, the biology of molecular hydrogen in animals has experiencing a surge of discoveries after Japanese scientists discovered that H_2 is a novel selective antioxidant in rats [9]. In contrast to the clear picture of H_2 in animals [10], the progress of H_2 functions in plants is just beginning [11]. Initially in plant biology, H_2 production is thought to be increased under various abiotic stresses and the normal growth conditions [12,13]. Afterwards, this gas was confirmed to be produced as an obligate by-product of the nitrogenase reaction [14], although other synthetic pathways of H_2 are still unknown.

Further experiments confirmed that exogenously applied with hydrogen-rich water (HRW) or the fumigation with H_2 could play vital roles in plant growth, development, and environmental responsiveness [15,16]. The degradation of fungicides remaining on the surface of plants could be accelerated by HRW through brassinosteroids signaling pathway, such as chlorothalonil (CHT) [17]. However, CHT was hardly absorbed by plants compared with CAR [18]. Genetic evidence showed that the expression of the *hydrogenase1* gene (*CrHYD1*) from *Chlamydomonas reinhardtii* not only increased endogenous H_2 production, but also confers Arabidopsis tolerance against salinity [19] and drought stress [20]. Additionally, H_2 could influence nitric oxide [21], abscisic acid [22], auxin [23], melatonin [19], and glutathione [24] signaling in plants.

This study wants to elucidate whether the application of H_2 could stimulate the degradation of CAR, a systemic fungicide applied in agriculture. First, genetic and physiological evidence showed that endogenous H_2 plays an important role in regulating CAR degradation in plant leaves through intensifying glutathione synthesis. Based on these results, we further suggest that CAR degradation might be controlled by genetic manipulation of endogenous H_2 or exogenously applied with H_2. Given the inherent toxicity of carbendazim and the intentional release into the environment [25], our results may contribute to the application of molecular hydrogen in agriculture and human health.

2. Results and Discussion

2.1. Exogenous H_2 Control of CAR Degradation

Though molecular hydrogen could enhance the metabolism of CHT in plants [17], the adsorption characteristics between CHT and CAR were entirely different. CHT was just attached to the leaf surface and hardly absorbed by plants, while CAR was absorbed into the plants and then exerted the medicinal effect. Thus, whether or how molecular hydrogen affects the metabolism of systemic fungicides in plants is still not clarified. To explore the problem above, tomato plants with six leaves were selected for the following experiments. Firstly, time curve analysis showed that after CAR addition, the absorption of this fungicide in tomato leaves was rapidly increased during the first 24 h period, followed by the progressive decline until 96 h (Figure S1). Therefore, tomato seedlings after treated with CAR for 24 h were used to assess the effect of exogenous molecular hydrogen on CAR degradation.

After the above CAR-treated seedlings were shifted to no CAR condition (CAR→dH_2O), its residue in leaves was moderately decreased during a 96-h period (Figure 1A). This result was consistent with the former findings [26], reflecting the photochemical catalytic degradation and biodegradation of CAR. Meanwhile, we observed that CAR addition for 24 h could rapidly stimulate the H_2 production, and reach the peak after 12 h (Figure 1B). These results demonstrated that the H_2 concentration was increased in response to CAR.

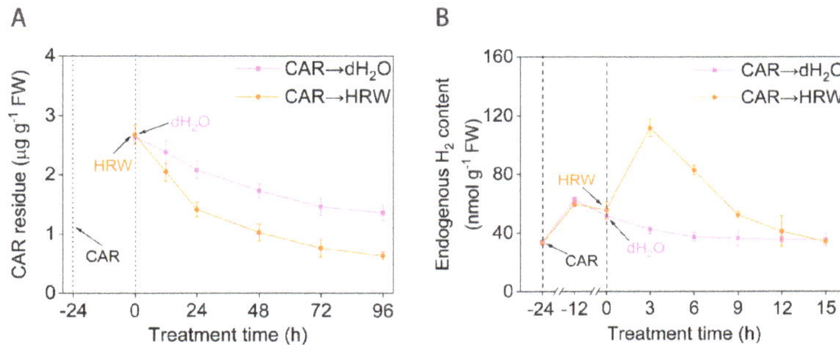

Figure 1. A possible link between the degradation of CAR and molecular hydrogen. Tomato seedlings at the six leaves stage were exposed to 10.46 mM CAR for 24 h, followed by the treatment with dH$_2$O and HRW for 24 h. Afterwards, time course of changes in CAR degradation curve (**A**) and endogenous H$_2$ contents (**B**) were determined. Error bars represent the standard deviation (SD; n = 3).

Subsequent experiments were performed to investigate whether this H$_2$ increase was directed towards counteracting CAR phytotoxicity. When treated with saturated HRW, which contained 0.78 mM H$_2$ produced by the electrolysis method, the degradation of CAR in tomato seedling leaves was remarkably intensified approximately in a time-dependent fashion (Figure 1A). For example, molecular hydrogen treated tomato leaves displayed residues of 86.2%, 68.1%, 59.6%, 52.2%, and 46.6% compared to those of the non-HRW treatment at 12, 24, 48, 72, and 96 h, respectively. Meanwhile, HRW addition could rapidly increase endogenous H$_2$ content in CAR-treated tomato leaves, reaching the peak after 3 h (Figure 1B). Similar results were discovered in seedlings of rice [27], alfalfa [28], and Arabidopsis [22] when treated in stressed conditions. Therefore, we further confirmed that endogenous molecular hydrogen metabolism might be regulated by environmental stimuli [29].

To assess the above results, six common crops that usually use CAR [30], including radish, cucumber, rice, rapeseed, alfalfa, and pepper, were selected for the further experiments. As shown in Figure S2, it was clearly illustrated that saturated HRW could remarkably promote the degradation of CAR in the above-mentioned crops, implying that molecular hydrogen control of CAR degradation might be a universal event.

2.2. Exogenous H$_2$ Did Not Alter the Antifungal Effect of CAR

Whether molecular hydrogen could influence the bactericidal effect of CAR is another concern. To answer this scientific question, *Alternaria solani*, which can cause several diseases on foliage, basal stems of seedlings, and fruits of tomato [31], were applied in PDA culture medium in the presence or absence of either HRW or CAR. After 10 days of inoculation, it was clearly observed that the colony diameter was differentially inhibited by CAR, when its concentrations were ranging from 2.62 to 20.92 mM (Figure 2A,B). Importantly, the above inhibition achieved by CAR was not altered regardless of HRW addition. Thus, HRW might not repress the growth and development of pathogenic bacteria. Consistently, no significant differences in the disease rate of leaf were observed in fungus-infected tomato seedlings at the six-leaves stage with or without HRW treatments (Figure 2C–F). The above results clearly showed that the antifungal effect of CAR was not altered by HRW. It was an interesting finding. This result further confirmed the possibility that the detoxification pathway triggered by molecular hydrogen might be mediated by a specific route, rather than binding tubulin molecule and inhibiting its role in microtubule assembly. Certainly, these parallel events should be carefully explored.

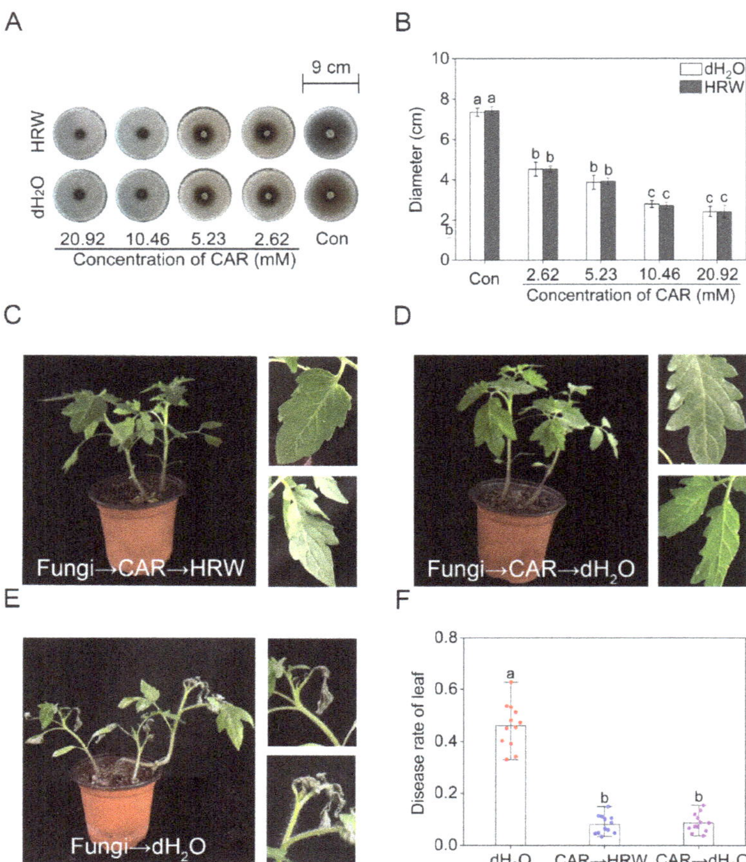

Figure 2. The antifungal effect of CAR was not affected by HRW. *Alternaria solani* was cultured in PDA culture mediums. After treatment for 10 days, the representative phenotypes were photographed (**A**). The diameter of the colony was counted (**B**). Tomato seedlings at the six–leaves stage were divided into three groups. The first and second groups were treated with 10.46 mM CAR for 24 h after inoculated by *Alternaria solani*. Afterwards, the two groups were respectively treated with HRW (CAR→HRW) or distilled water (CAR→dH$_2$O) every two days. The third group was inoculated with *Alternaria solani* for 24 h and treated with distilled water (dH$_2$O) every two days. After 15 days, the representative phenotypes were photographed (**C–E**). The diseased leaves rate of three different treatments was counted (**F**). The error bars represent the SDs (n = 3 for (**B**), n = 12 for (**F**)), and the different letters indicate significantly different values ($p < 0.05$ according to Tukey's multiple test).

2.3. Glutathione Involvement in Molecular Hydrogen Control of CAR Degradation

Since glutathione (GSH) is identified as an important detoxification compound and defensive signaling molecule in both human disease responses and plant adaptation to environmental stresses [32], the changes in GSH-metabolism and detoxification genes expression were analyzed by using RT-qPCR. As shown in Figure 3, CAR stimulated the expression of *GS* (encoding glutathione synthetase), *GST* genes (encoding glutathione *S*-transferase), *GPX* (encoding glutathione synthetase), *ABC2* (encoding ABC transporters), and *CYP724B2* (encoding one of the members of brassinosteroids) in the plants [33,34]. It could be speculated that the enhancement of the above gene expression following exposure to CAR might be an initial defense response for intensifying the GSH-metabolism and detoxification systems. Afterwards, we observed that HRW addition could further stimulate the expression of the above genes compared to the dH_2O control samples.

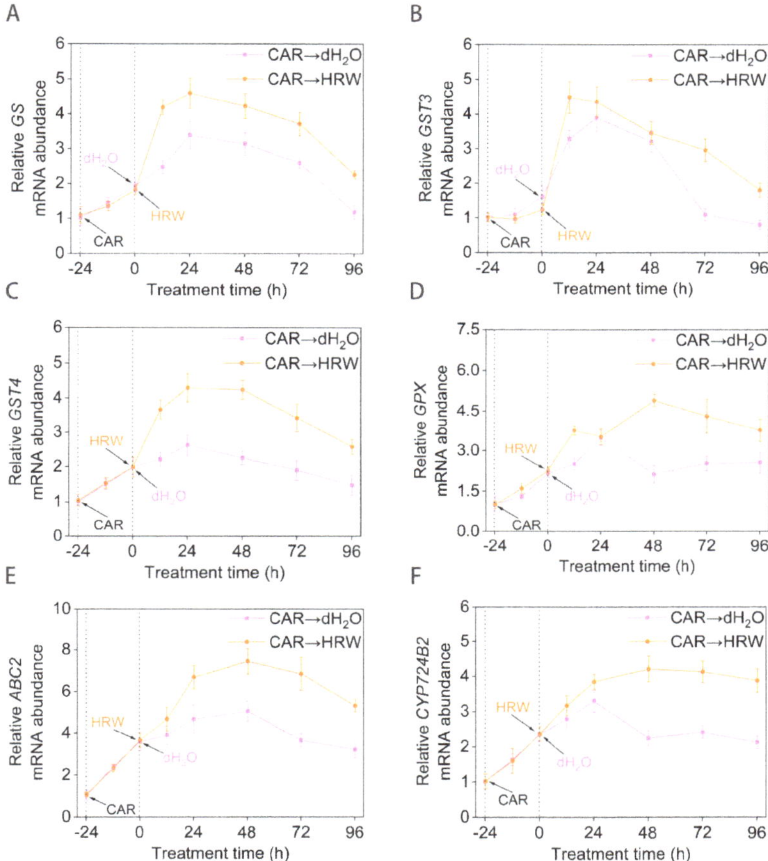

Figure 3. Time curve of GSH−metabolism and detoxification genes expression. Tomato seedlings at the six leaves stage were exposed to 10.46 mM CAR for 24 h, followed by the treatment with dH_2O or HRW for 24 h. Afterwards, the time course of changes in *GS* (**A**), *GST3* (**B**), *GST4* (**C**), *GPX* (**D**), *ABC2* (**E**), *CYP724B2* (**F**) genes expression was determined. Error bars represent the standard deviation (SD; n = 3).

Subsequently, endogenous GSH levels were monitored under the identical conditions. Although the pretreatment with CAR for 24 h could not obviously influence the changes of reduced GSH levels [34], the addition of HRW did trigger GSH production, peaking at 24 h after treatment, then followed by a gradual decrease, but still above the basal level (Figure 4A). Unlike the responses in reduced GSH, oxidized GSH (GSSG) production in tomato seedling leaves was stimulated by CAR, which was further maintained by HRW up to 48 h after treatment compared to non-HRW control samples (Figure 4B). A similar change in GSH pool, especially GSSG, achieved by CAR, was previously reported in tomato plants [26]. The above results clearly showed that an increase in the concentration of GSH might be one of the earliest responses participating in the signaling transduction triggered by molecular hydrogen in tomato leaves. Subsequent results revealed that GSH addition could improve photosynthetic characteristics. As expected, treatment with CAR could severely inhibit the values of the net photosynthetic rate (Pn), the maximum PSII quantum yield (Fv/Fm), and photochemical quenching coefficient (qP), reflecting the phytotoxicity of CAR (Figure S3A–C), which was also accompanied with the increased lipid peroxidation in tomato seedling leaves (TBARS; Figure S3D). After GSH addition, the above parameters could be differentially improved, especially with 5 mM GSH exhibiting the maximal responses, which was used subsequently.

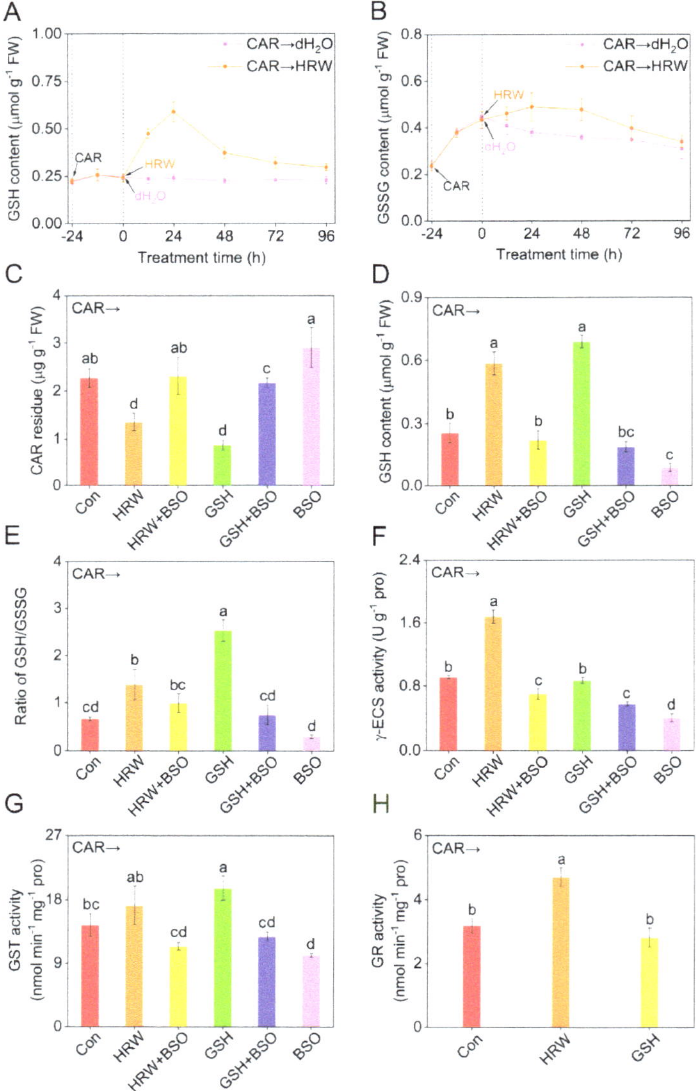

Figure 4. Changes in glutathione−related metabolism and CAR residues. Seedlings at the six−leaves stage were exposed to 10.46 mM CAR for 24 h, followed by the treatment with dH_2O (Con), HRW, GSH, or BSO, alone or their combinations. Afterwards, time−course changes in endogenous reduced GSH (**A**) and GSSG (**B**) contents were analyzed. After treatments for 24 h, CAR residue (**C**), reduced GSH content (**D**), the ratio of GSH/GSSG (**E**), and γ-ECS (**F**), GST (**G**), and GR (**H**) activities in tomato leaves were determined. Error bars represent the standard deviation (SD; n = 3). Bars with different letters are significantly different ($p < 0.05$) according to Tukey's multiple test.

To assess the contribution of endogenous GSH in the response of CAR degradation achieved by molecular hydrogen, $_L$-buthionine-sulfoximine (BSO), an effective inhibitor of γ-glutamylcysteine synthetase (γ-ECS) responsible for GSH synthesis [35,36], was applied individually or simultaneously. Further experiments showed an obvious increase of endogenous GSH content, and importantly, a significant reduction in CAR content, after the addition of GSH (Figure 4C,D), mimicking the responses of HRW. By contrast, the above-mentioned results were remarkably abolished by BSO, reflecting the important function of endogenous GSH. Alone, BSO could result in the accumulation of CAR and in the depletion of GSH.

The GSH biosynthesis induction by molecular hydrogen was previously observed in the stressed conditions [37] or during lateral root development [24]. To further decipher the protective role of GSH in molecular hydrogen control of CAR degradation, GSH metabolism was examined. As shown in Figure 4E, in the presence of CAR, both HRW and especially GSH could significantly enhance the ratio of GSH/GSSG. Unlike the effect of exogenous GSH, the above response of HRW might be caused by the increased activity of γ-ECS (Figure 4F), an important enzyme catalyzed GSH synthesis [36]. These indicated the stimulation of GSH production achieved by molecular hydrogen via intensifying γ-ECS activity, and similar results were found in both animals and plants, especially under environmental stimuli [37]. Consistently, the above responses in GSH/GSSG and γ-ECS activity conferred by HRW were abolished by the removal of endogenous GSH in the presence of BSO, reflecting that the HRW control of the response is γ-ECS-dependent. When applied alone, BSO did result in negative effects.

Both glutathione *S*-transferase (GST) and glutathione reductase (GR) are two enzymes related to GSH metabolism. Between these, GR is responsible for reducing GSSG back to GSH [38]. Particularly, ample evidence showed that GSH control of organic compounds detoxification is closely associated with the reaction catalyzed by GST, and its product, GSH conjugates, could be transported into the vacuole [39]. Accordingly, the changes in GR and GST were investigated and compared in the presence or absence of HRW or GSH, with or without the addition of BSO.

For GST, further results clearly revealed that both HRW and GSH intensified its activities, which were further impaired by BSO (Figure 4G). Importantly, these data were correlated to changes in CAR residues (Figure 4C). These might be explained by the fact that CAR residues might be bound to GSH catalyzed by enhanced GST activity to form less reactive and toxic conjugates for subsequent transportation and degradation [40]. Similar changes were observed in the contents of nonprotein thiols (NPT; Figure S4A). Since NPT were closely associated with GSH conjugates, we further deduced that the stimulation of NPT synthesis achieved by molecular hydrogen might reflect the decomposition products of pesticide to some extent [41]. After the addition of CAR, unlike the response of GSH, HRW could remarkably increase the activity of GR (Figure 4H). These changes also matched with the GSH content and the ratio of GSH and GSSG, further supporting the important function of endogenous GSH in the molecular hydrogen control of CAR degradation.

2.4. Transcriptional Regulation of GSH-Metabolism and Detoxification Genes

To probe the molecular mechanism underlying molecular hydrogen control of GSH synthesis, RT-qPCR was applied. As shown in Figure 5A–C, HRW strongly stimulated the expression of *GR*, *GSH1* (encoding γ-ECS) [42], and *GST2* in the presence of CAR, all of which were matched with the increased activities of GR (Figure 4H), γ-ECS (Figure 4F), and GST (Figure 4G), thus promoting endogenous GSH production in CAR-treated tomato seedling leaves (Figure 4A).

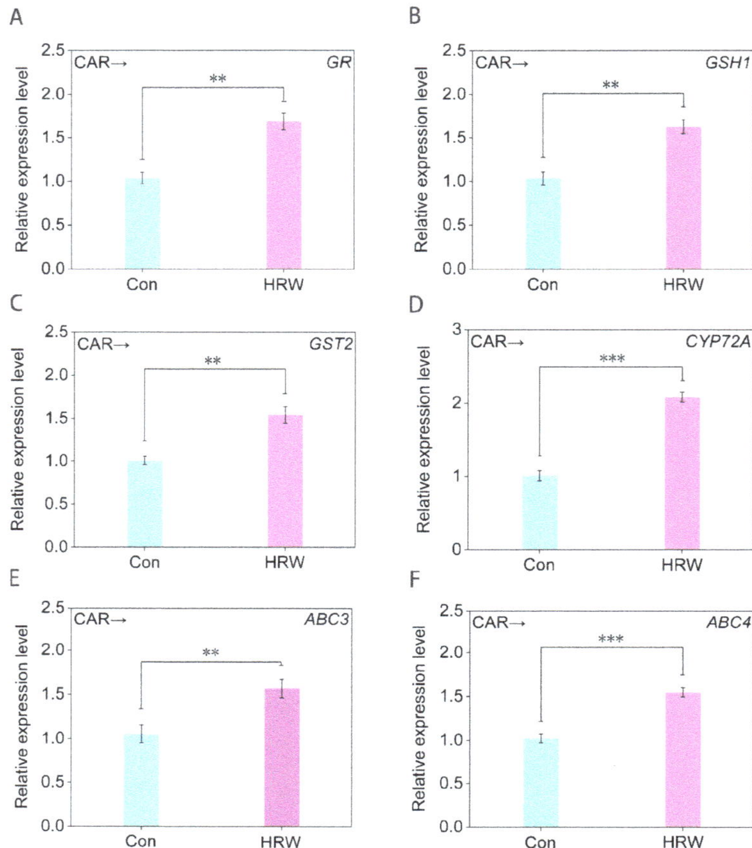

Figure 5. Changes in the transcription levels of GSH−dependent genes and detoxification genes. Tomato seedlings at the six−leaves stage were treated with 10.46 mM CAR for 24 h, followed by treatment with dH$_2$O (Con) or HRW. After 24 h of treatment, *GR* (**A**), *GSH1* (**B**), *GST2* (**C**), *CYP72A* (**D**), *ABC3* (**E**), and *ABC4* (**F**) transcript levels were analyzed by RT-qPCR. Error bars represent the standard deviation (SD; n = 3). ** and *** indicate significant difference results at $p < 0.01$ or $p < 0.001$ analyzed by independent-sample *t*-test.

Meanwhile, three detoxification genes, including *CYP72A*, *ABC3*, and *ABC4*, were selected to evaluate their involvement of exogenous H$_2$ control of CAR degradation. Among these, *CYP72A* encodes one of the members of cytochrome P450s, which are indispensable for hormone synthesis and detoxification in both animals and plants [43]. ATP-binding cassette (ABC) transporters also participated in the detoxification of environmental pollutants [44,45]. Similarly, we further observed that in the presence of CAR, the expression of *CYP72A* (Figure 5D), *ABC3* (Figure 5E), and *ABC4* (Figure 5F) were significantly intensified by HRW, which were also correlated with the increased degradation of CAR (Figure 1A).

2.5. CAR-Triggered Redox Imbalance Was Abolished by HRW

For pesticide-induced phytotoxicity, ample evidence revealed that insecticides could influence redox balance, including resulting in reactive oxygen species (ROS) production and oxidative damage, in both animal and plant cells [46]. RT-qPCR results showed that *CAT1* (encoding catalase), *G-POD* (encoding peroxidase), *APX1* (encoding ascorbate peroxidase), *Cu/Zn-SOD* (encoding superoxide dismutase), *MDHAR* (encoding monodehydroascorbate reductase), and *DHAR* (encoding dehydroascorbate reductase) genes (Zhou et al., 2017) were up-regulated after applying CAR (Figure S5). Furthermore, the combined treatment of CAR and HRW induced the transcription of the above-mentioned genes more strongly than CAR or HRW alone, indicating an additive effect of CAR and HRW on the induction.

Subsequent results showed that in the presence of either HRW or GSH, both H_2O_2 and O_2^- distribution (two important components of ROS; Figure S4B,C) and TBARS accumulation (an index related to oxidative damage; Figure S4D) were found to be abolished in CAR-treated tomato leaves, indicating the reconstructing redox homeostasis achieved by both HRW and GSH. These effects were sensitive to the removal of endogenous GSH by the addition of BSO, thus further emphasizing the participation of GSH. Similar tendencies were found in the changes of four antioxidant enzymes activities, including superoxide dismutase (SOD; Figure S4E), catalase (CAT; Figure S4F), ascorbate peroxidase (APX; Figure S4G), and guaiacol peroxidase (POD; Figure S4H).

2.6. Genetic Evidence Revealed That Endogenous Molecular Hydrogen Can Positively Influence Carbendazim Degradation via GSH

Exogenously applied molecular hydrogen may not faithfully mimic the function of endogenous H_2 [22]. Thus, the genetic-based approach, using genetic materials with altered endogenous H_2 levels, is likely to be more accurate for dissecting the nature function of molecular hydrogen, although its metabolism is still not fully elucidated in plants [11].

In the previous reports, the expression of *hydrogenase1* gene from *Chlamydomonas reinhardtii* (*CrHYD1*) in Arabidopsis represents an interesting method to assess the functions of endogenous production of H_2 in plant cells, since molecular hydrogen derived from *CrHYD1* expression could modify the stomatal closure [20] and improve salinity tolerance [19]. Here, compared to those in the wild-type (WT) with or without being transformed with the empty vector (EV), six *CrHYD1* lines not only showed increased H_2 production in the presence of CAR (Figure S6), but importantly, displayed the lower residues of CAR in leaves, especially *CrHYD1-3*, *CrHYD1-5*, and *CrHYD1-6* (Figure 6A), and similar beneficial roles of endogenous H_2 were discovered according to the changes in stomatal bioassays [20] and salinity phenotypic analysis [19]. Under the identical treatments, contrasting changes in endogenous GSH were also observed (Figure 6B), thus reflecting the possible negative correlation between CAR residues and GSH contents. This deduction was further supported by the ones exogenously applied with either GSH or BSO after the spraying with CAR in the above transgenic lines and wild-type plants. For example, after the addition of GSH, the decreased CAR residues and increased GSH contents observed in transgenic lines were differentially intensified. Contrasting results were found when BSO was applied, which could inhibit endogenous GSH biosynthesis. The above results clearly provided genetic evidence, showing that endogenous H_2 is an endogenous regulator for CAR degradation in a GSH-dependent fashion.

In summary, our results clearly demonstrated that either exogenously applied or endogenously increased molecular hydrogen can decrease CAR residues in plant leaves. The present findings also indicated the possible role of GSH-dependent pathway in the detoxification of CAR, although a corresponding mechanism has not been fully elucidated.

Although several reports discovered that both brassinosteroids [40] and melatonin [26] could help plants degrade pesticides, the application with HRW without using the above conventional chemical additives might be more environmentally friendly for sustainable agriculture. More importantly, this study opened a new way for molecular hydrogen application to degrade the systemic fungicide CAR in a crop.

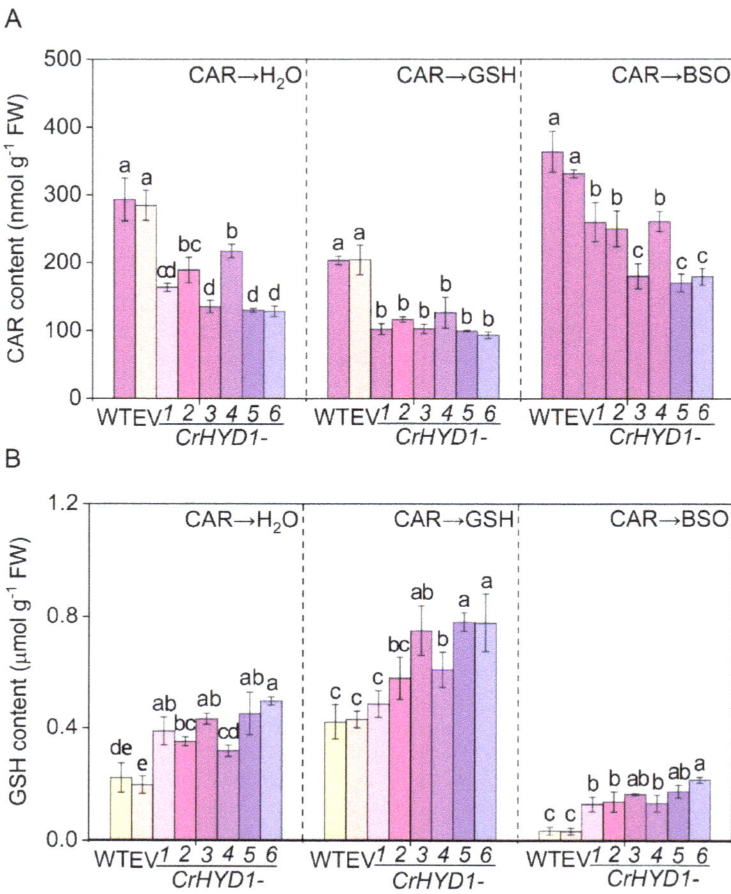

Figure 6. Genetic evidence showing the important role of endogenous molecular hydrogen in promoting the degradation of CAR via GSH synthesis. After *Arabidopsis* seedlings were sprayed with 10.46 mM CAR for 24 h, seedlings were treated with dH$_2$O (Con), GSH, or BSO for 24 h. Afterwards, CAR residues (**A**) and GSH content (**B**) were measured. The error bars represent the SDs (n = 3; 20 plants/treatment/repeat). Different letters indicate significantly different values ($p < 0.05$ according to Tukey's multiple test).

3. Materials and Methods

3.1. Chemicals

DL-Buthionine-*S* (BSO; CAS 83730-53-4; purity $\geq$ 97%) and reduced glutathione (GSH; CAS 27025-41-8; purity $\geq$ 98%) were purchased from Sigma-Aldrich (St. Louis, MO, USA). Carbendazim (CAR; 50% active ingredient) was obtained from Yingkouleike (Liaoning, China).

3.2. Plant Material, Growth Conditions, and Treatments

According to the previous methods [47], five-day-old tomato (*Solanum lycopersicum* L. cv. baiguoqiangfeng) seedlings were transferred to a flowerpot containing a mixture of peat and vermiculite (3:1; v/v) and were cultivated in a growth chamber with a light intensity of 200 μmol m^{-2} s^{-1} under a light cycle of 14 h of light and 10 h of dark at 24 °C. Full-strength Hoagland's nutrient solution was replaced every two days until six true leaves were grown.

The *CrHYD1* transgenic strain expressed the *hydrogenase1* gene from *C. reinhardtii* in Arabidopsis plants under the control of the cauliflower mosaic virus (*CaMV*) 35S promoter [20]. Seven-day-old Arabidopsis seedlings including wild-type, control plants transformed with the empty vector (EV), and six *CrHYD1* transgenic lines (*CrHYD1-1*, *CrHYD1-2*, *CrHYD1-3*, *CrHYD1-4*, *CrHYD1-5*, and *CrHYD1-6*) were cultivated in a growth chamber with a light intensity of 150 μmol m^{-2} s^{-1} under a light cycle of 16 h of light and 8 h of dark at 23 °C for 6 weeks.

In order to elucidate the interaction between molecular hydrogen and CAR degradation (Figure 1) and whether GSH metabolism is stimulated by molecular hydrogen (Figure 4), CAR was dissolved in distilled water to prepare a solution of 10.46 mM (the concentration used in practical applications) and further sprayed on tomato leaves [34], with 20 mL CAR used for each potted plant. After 24 h treatment, the above potted plants were divided into two groups and placed in two identical trays (20–40 pots for each). The two trays were filled with 500 mL dH$_2$O or HRW, respectively.

In order to assess whether the CAR degradation achieved by molecular hydrogen was related to GSH metabolism (Figures 4 and 6), or altered the redox balance (Figure S4), 10.46 mM CAR was used, with 20 mL CAR sprayed for each potted plant. After 24 h of treatment, the above potted plants were divided into six groups, and respectively placed in six identical trays (20–40 pots for each).

Four groups were respectively filled with 500 mL distilled water, and each potted plant was sprayed with 5 mL distilled water containing an equal ratio of organic solvent (CAR→dH$_2$O/Con), or sprayed with 5 mL 5 mM GSH (CAR→GSH), or sprayed with 5 mL 5 mM GSH and 5 mL 1 mM BSO (CAR→GSH + BSO), or sprayed with 5 mL 1 mM BSO (CAR→BSO).

Two other groups were respectively filled with 500 mL HRW, and each potted plant was sprayed with 5 mL distilled water containing an equal ratio of organic solvent (CAR→HRW), or sprayed with 5 mL 1 mM BSO (CAR→HRW + BSO).

3.3. Preparation of Hydrogen-Rich Water

According to the previous method [22], the saturated hydrogen-rich water (HRW) was prepared by using a SHC-300 H$_2$ generator (Saikesaisi Hydrogen Energy Co., Ltd., Jinan, China). The H$_2$ concentration in the above HRW was about 0.78 mM, and this saturated solution was applied subsequently.

3.4. Determination of H$_2$ Content

Endogenous H$_2$ content was measured with headspace gas chromatography (Tianmei GC7900 equipped with a thermal conductivity detector, Tianmei Scientific Instrument Co., Ltd., Shanghai, China) as described previously [48].

3.5. Determination of GSH Content by UPLC Analysis

According to previous methods [49], the GSH contents in leaves were determined with ultra-high performance liquid chromatography (UPLC; Agilent 1290, Agilent, Palo Alto, CA, USA). Total GSH and GSSG contents were determined by UPLC, and reduced GSH content was estimated from the difference between total GSH and GSSG.

3.6. Determination of CAR Residues in Tomato Leaves

CAR residues were determined by UPLC (Agilent 1290, Agilent, Palo Alto, CA, USA) [26]. The wavelength of detection was 280 nm, and the injection volume was 2 μL.

3.7. Indoor Toxicity Assessment and Experiment of Early Blight Resistance

According to a previous method [50], indoor toxicity has been determined. *Alternaria solani* (Shanghai Bioresource Collection Center, SHBCC, Shanghai, China) was inoculated into the center of potato dextrose agar (PDA) mediums containing 0, 2.62, 5.23, 10.46, or 20.92 mM CAR at 28 °C. Organic membrane filters with the pore size of 0.22 μm were used to filter HRW and distilled water, which were added to the culture medium. The colony diameter was determined after the incubation for 10 days.

After *Alternaria solani* was propagated on PDA mediums for 10 days, 2 mL of sterile water was added, and the spore was slowly scraped with a microscope slide. After inoculation, the spore concentration of the suspension was adjusted to 1×10^4 conidia mL^{-1}, which was sprayed on tomato seedlings at the six-leaves stage [51]. The disease rate of leaves was calculated after 15 d of treatment [52].

3.8. Analysis of the GSH Cycle

According to the previous method [53], the activity of GR (glutathione reductase) was analyzed by detecting the rate of decrease in the absorbance of 340 nm. GST (glutathione *S*-transferase) and γ-ECS (γ-glutamylcysteine synthetase) activities were assayed with GST and γ-ECS activity kit (Nanjing JianCheng Bioengineering Institute, Nanjing, China). One unit (U) of γ-ECS activity was defined as the production of 1 µmol inorganic phosphorus per milligram of tissue protein per hour.

3.9. RT-qPCR Determination of Transcript Levels of Genes

After the isolation of total RNA and cDNA synthesis, real-time quantitative PCR (RT-qPCR) was performed [54], and the gene-specific primers were listed in Table S1. *GADPH* was used as an internal control gene. Three technical replicates of RT-qPCR were performed per gene-specific primers pair.

3.10. Statistical Analysis

Statistical analysis was performed using IBM SPSS Statistics 16.0 software. Statistical significance was analyzed by ANOVA analysis in combination with Tukey's multiple test ($p < 0.05$) or independent-sample *t*-test. ** and *** indicate significant difference results at $p < 0.01$ or $p < 0.001$.

4. Conclusions

The above molecular and genetic evidence demonstrated that H_2 could promote the CAR degradation in plants. GSH operates a downstream molecule functioning in the above process. In addition, molecular hydrogen could effectively enhance plant antioxidant capacity, thus reestablishing redox balance. Importantly, the results from this study will assist and promote further efforts to bring the findings of basic hydrogen biology research to hydrogen-based agriculture, thus meeting the dietary needs of 7.6 billion people in a healthy and sustainable manner.

Supplementary Materials: The following supporting information can be downloaded at: https://www.mdpi.com/article/10.3390/plants11050621/s1, Table S1: The sequences primer sequences for qPCR; Figure S1: Time curve of CAR absorption in tomato leaves; Figure S2: Molecular hydrogen-decreased CAR residues might be a universal event; Figure S3: GSH could alleviate the photoinhibition of CAR on tomato leaves; Figure S4: Changes in redox homeostasis; Figure S5: Comparison of expression of antioxidant genes; Figure S6: H2 production in WT, EV, and six CrHYD1 lines. Non-protein thiol content was determined as previously described [55]. The photosynthetic parameters of tomato leaves were determined according to the method [21,56,57]. Antioxidant index of tomato leaves was determined according to the method [28,58–60].

Author Contributions: T.Z.: writing—original draft, methodology, formal analysis. Y.W.: methodology, formal analysis. Z.Z.: data curation. S.X.: methodology. W.S.: formal analysis, supervision, writing—review and editing. All authors have read and agreed to the published version of the manuscript.

Funding: This work was supported by the Funding from Center of Hydrogen Science, Shanghai Jiao Tong University, and Foshan Agriculture Science and Technology Project (Foshan City Budget No. 140, 2019.)

Institutional Review Board Statement: Not applicable.

Informed Consent Statement: Not applicable.

Data Availability Statement: Data is contained within the article or supplementary material.

Conflicts of Interest: The authors declare that they have no known competing financial interest or personal relationships that could have appeared to influence the work reported in this paper.

References

1. Oerke, E.C. Crop losses to pests. *J. Agric. Sci.* **2006**, *144*, 31–43. [CrossRef]
2. Lee, S.J.; Mehler, L.; Beckman, J.; Diebolt-Brown, B.; Prado, J.; Lackovic, M.; Waltz, J.; Mulay, P.; Schwartz, A.; Mitchell, Y.; et al. Acute pesticide illnesses associated with off-target pesticide drift from agricultural applications: 11 states, 1998–2006. *Environ. Health Perspect.* **2011**, *119*, 1162–1169. [CrossRef]
3. European Food Safety Authority (EFSA). Conclusion on the peer review of the pesticide risk assessment of the active substance carbendazim. *EFSA J.* **2010**, *8*, 1598. [CrossRef]
4. Davidse, L.C. Benzimidazole fungicides: Mechanism of action and biological impact. *Annu. Rev. Phytopathol.* **1986**, *24*, 43–65. [CrossRef]
5. Zhang, X.; Huang, Y.; Harvey, P.R.; Li, H.; Ren, Y.; Li, J.; Wang, J.; Yang, H. Isolation and characterization of carbendazim-degrading *Rhodococcus erythropolis* djl-11. *PLoS ONE* **2013**, *8*, e74810. [CrossRef]
6. Singh, S.; Singh, N.; Kumar, V.; Datta, S.; Wani, A.B.; Singh, D.; Singh, K.; Singh, J. Toxicity, monitoring and biodegradation of the fungicide carbendazim. *Environ. Chem. Lett.* **2016**, *14*, 317–329. [CrossRef]
7. Besson-Bard, A.; Courtoisa, C.; Gauthiera, A.; Dahana, J.; Dobrowolskab, G.; Jeandrozc, S.; Pugin, A.; Wendehenne, D. Nitric oxide in plants: Production and cross-talk with Ca^{2+} signaling. *Mol. Plant* **2008**, *1*, 218–228. [CrossRef]
8. Yin, Y.; Zhou, Y.; Zhou, Y.; Shi, K.; Zhou, J.; Yu, Y.; Yu, J.; Xia, X. Interplay between mitogen-activated protein kinase and nitric oxide in brassinosteroid-induced pesticide metabolism in *Solanum lycopersicum*. *J. Hazard. Mater.* **2016**, *316*, 221–231. [CrossRef]
9. Ohsawa, I.; Ishikawa, M.; Takahashi, K.; Watanabe, M.; Nishimaki, K.; Yamagata, K.; Katsura, K.; Katayama, Y.; Asoh, S.; Ohta, S. Hydrogen acts as a therapeutic antioxidant by selectively reducing cytotoxic oxygen radicals. *Nat. Med.* **2007**, *13*, 688–694. [CrossRef]
10. Ohta, S. Molecular hydrogen as a preventive and therapeutic medical gas: Initiation, development and potential of hydrogen medicine. *Pharmacol. Ther.* **2014**, *144*, 1–11. [CrossRef]
11. Shen, W.; Sun, X.J. Hydrogen biology: It is just beginning. *Chin. J. Biochem. Mol. Biol.* **2019**, *35*, 1037–1050. [CrossRef]
12. Gaffron, H. Reduction of carbon dioxide with molecular hydrogen in green algae. *Nature* **1939**, *143*, 204–205. [CrossRef]
13. Renwick, G.M.; Giumarro, C.; Siegel, S.M. Hydrogen metabolism in higher plants. *Plant Physiol.* **1964**, *39*, 303–306. [CrossRef] [PubMed]
14. Hunt, S.; Layzell, D.B. Gas-exchange of legume nodules and the regulation of nitrogenase activity. *Annu. Rev. Plant Physiol. Plant Mol. Biol.* **1993**, *44*, 483–511. [CrossRef]
15. Wang, Y.Q.; Liu, Y.H.; Wang, S.; Du, H.M.; Shen, W. Hydrogen agronomy: Research progress and prospects. *J. Zhejiang Univ. Sci. B* **2020**, *21*, 841–855. [CrossRef]
16. Li, L.; Lou, W.; Kong, L.; Shen, W. Hydrogen commonly applicable from medicine to agriculture: From molecular mechanisms to the field. *Curr. Pharm. Des.* **2021**, *27*, 747–759. [CrossRef]
17. Wang, Y.; Zhang, T.; Wang, J.; Xu, S.; Shen, W. Regulation of chlorothalonil degradation by molecular hydrogen. *J. Hazard. Mater.* **2022**, *424*, 127291. [CrossRef]
18. Wang, Z.W.; Huang, J.; Chen, J.Y.; Li, F.L. Time-dependent movement and distribution of chlorothalonil and chlorpyrifos in tomatoes. *Ecotoxicol. Environ. Saf.* **2013**, *93*, 107–111. [CrossRef]
19. Su, J.; Yang, X.; Shao, Y.; Chen, Z.; Shen, W. Molecular hydrogen–induced salinity tolerance requires melatonin signaling in *Arabidopsis thaliana*. *Plant Cell Environ.* **2021**, *44*, 476–490. [CrossRef]
20. Zhang, Y.; Cheng, P.; Wang, Y.; Li, Y.; Su, J.; Chen, Z.; Yu, X.; Shen, W. Genetic elucidation of hydrogen signaling in plant osmotic tolerance and stomatal closure via hydrogen sulfide. *Free Radic. Biol. Med.* **2020**, *161*, 1–14. [CrossRef]
21. Su, J.; Zhang, Y.; Nie, Y.; Cheng, D.; Wang, R.; Hu, H.; Chen, J.; Zhang, J.; Du, Y.; Shen, W. Hydrogen-induced osmotic tolerance isassociated with nitric oxide-mediated proline accumulation and reestablishment of redox balance in alfalfa seedlings. *Environ. Exp. Bot.* **2018**, *147*, 249–260. [CrossRef]
22. Xie, Y.; Mao, Y.; Zhang, W.; Lai, D.; Wang, Q.; Shen, W. Reactive oxygen species-dependent nitric oxide production contributes to hydrogen-promoted stomatal closure in Arabidopsis. *Plant Physiol.* **2014**, *165*, 759–773. [CrossRef]
23. Cao, Z.; Duan, X.; Yao, P.; Cui, W.; Cheng, D.; Zhang, J.; Jin, Q.; Chen, J.; Dai, T.; Shen, W. Hydrogen gas is involved in auxin-induced lateral root formation by modulating nitric oxide synthesis. *Int. J. Mol. Sci.* **2017**, *18*, 2084. [CrossRef] [PubMed]
24. Liu, F.; Lou, W.; Wang, J.; Li, Q.; Shen, W. Glutathione produced by γ-glutamyl cysteine synthetase acts downstream of hydrogen to positively influence lateral root branching. *Plant Physiol. Biochem.* **2021**, *167*, 68–76. [CrossRef] [PubMed]
25. De Wilde, T.; Spanoghe, P.; Ryckeboer, J.; Jaeken, P.; Springael, D. Sorption characteristics of pesticides on matrix substrates used in biopurification systems. *Chemosphere* **2009**, *75*, 100–108. [CrossRef]

26. Yan, Y.; Sun, S.; Zhao, N.; Yang, W.; Shi, Q.; Gong, B. *COMT1* overexpression resulting in increased melatonin biosynthesis contributes to the alleviation of carbendazim phytotoxicity and residues in tomato plants. *Environ. Pollut.* **2019**, *252*, 51–61. [CrossRef]
27. Xu, S.; Jiang, Y.; Cui, W.; Jin, Q.; Zhang, Y.; Bu, D.; Fu, J.; Wang, R.; Zhou, F.; Shen, W. Hydrogen enhances adaptation of rice seedlings to cold stress via the reestablishment of redox homeostasis mediated by miRNA expression. *Plant Soil* **2017**, *414*, 53–67. [CrossRef]
28. Jin, Q.; Zhu, K.; Cui, W.; Xie, Y.; Han, B.; Shen, W. Hydrogen gas acts as a novel bioactive molecule in enhancing plant tolerance to paraquat-induced oxidative stress via the modulation of heme oxygenase-1 signalling system. *Plant Cell Environ.* **2013**, *36*, 956–969. [CrossRef]
29. Zulfqar, F.; Russell, G.; Hancock, J.T. Molecular hydrogen in agriculture. *Planta* **2021**, *254*, 56. [CrossRef]
30. Xu, X.; Chen, J.; Li, B.; Tang, L. Carbendazim residues in vegetables in China between 2014 and 2016 and a chronic carbendazim exposure risk assessment. *Food Control* **2018**, *91*, 20–25. [CrossRef]
31. Chaerani, R.; Voorrips, R.E. Tomato early blight (*Alternaria solani*): The pathogen, genetics, and breeding for resistance. *J. Gen. Plant Pathol.* **2006**, *72*, 335–347. [CrossRef]
32. Coleman, J.O.D.; Randall, R.; Blake-Kalff, M.M.A. Detoxification of xenobiotics in plant cells by glutathione conjugation and vacuolar compartmenttaiization: A fluorescent assay using monochiorobimane. *Plant Cell Environ.* **1997**, *20*, 449–460. [CrossRef]
33. Yu, G.B.; Zhang, Y.; Ahammeda, G.J.; Xia, X.J.; Mao, W.H.; Shi, K.; Zhou, Y.H.; Yu, J.Q. Glutathione biosynthesis and regeneration play an important role in the metabolism of chlorothalonil in tomato. *Chemosphere* **2013**, *90*, 2563–2570. [CrossRef] [PubMed]
34. Wang, J.; Jiang, Y.; Chen, S.; Xia, X.; Shi, K.; Zhou, Y.; Yu, Y.; Yu, J. The different responses of glutathione-dependent detoxification pathway to fungicide chlorothalonil and carbendazim in tomato leaves. *Chemosphere* **2010**, *79*, 958–965. [CrossRef] [PubMed]
35. Seyfried, J.; Soldner, F.; Schulz, J.B.; Klockgether, T.; Kovar, K.A.; Wüllner, U. Differential effects of ₗ-buthionine sulfoximine and ethacrynic acid on glutathione levels and mitochondrial function in PC12 cells. *Neurosci. Lett.* **1999**, *264*, 1–4. [CrossRef]
36. Vernoux, T.; Wilson, R.C.; Seeley, K.A.; Reichheld, J.P.; Muroy, S.; Brown, S.; Maughan, S.C.; Cobbett, C.S.; Montagu, M.V.; Inzé, D.; et al. The *ROOT MERISTEMLESS1/CADMIUM SENSITIVE2* Gene defines a glutathione-dependent pathway involved in initiation and maintenance of cell division during postembryonic root development. *Plant Cell* **2000**, *12*, 97–109. [CrossRef]
37. Cui, W.; Gao, C.; Fang, P.; Lin, G.; Shen, W. Alleviation of cadmium toxicity in *Medicago sativa* by hydrogen-rich water. *J. Hazard. Mater.* **2013**, *260*, 715–724. [CrossRef]
38. Noctor, G.; Mhamdi, A.; Chaouch, S.; Han, Y.; Neukermans, J.; Marquez-Garcia, B.; Queval, G.; Foyer, C.H. Glutathione in plants: An integrated overview. *Plant Cell Environ.* **2012**, *35*, 454–484. [CrossRef]
39. Marrs, K.A. The functions and regulation of glutathione S-transferases in plants. *Annu. Rev. Plant Physiol. Plant Mol. Biol.* **1996**, *47*, 127–158. [CrossRef]
40. Xia, X.; Zhang, Y.; Wu, J.; Wang, J.; Zhou, Y.; Shi, K.; Yu, Y.; Yu, J. Brassinosteroids promote metabolism of pesticides in cucumber. *J. Agric. Food Chem.* **2009**, *57*, 8406–8413. [CrossRef]
41. Yu, G.; Wei, J.; Chen, X.; Li, X.; Liu, X.; Ye, X.; Zhang, N.; Sun, W. The effect of glutathione in the regulation of the degradation of residual fungicide in tomato. *Int. J. Agric. Biol.* **2018**, *20*, 1873–1879. [CrossRef]
42. Cobbett, C.S.; May, M.J.; Howden, R.; Rolls, B. The glutathione-deficient, cadmium-sensitive mutant, *cad2–1*, of *Arabidopsis thaliana* is deficient in γ-glutamylcysteine synthetase. *Plant J.* **1998**, *16*, 73–78. [CrossRef] [PubMed]
43. Religia, P.; Nguyen, N.D.; Nong, Q.D.; Matsuura, T.; Kato, Y.; Watanabe, H. Mutation of the cytochrome P450 *CYP360A8* gene increases sensitivity to paraquat in *Daphnia magn*. *Environ. Toxicol. Chem.* **2021**, *40*, 1279–1288. [CrossRef]
44. Deen, M.; Vries, E.G.; Timens, W.; Scheper, R.J.; Timmer-Bosscha, H.; Postma, D.S. ATP-binding cassette (ABC) transporters in normal and pathological lung. *Resp. Res.* **2005**, *6*, 59. [CrossRef] [PubMed]
45. Rea, R.A. Plant ATP-binding cassette transporters. *Annu. Rev. Plant Biol.* **2007**, *58*, 347–375. [CrossRef] [PubMed]
46. Ahammed, G.J.; Ruan, Y.P.; Zhou, J.; Xia, X.J.; Shi, K.; Zhou, Y.H.; Yu, J.Q. Brassinosteroid alleviates polychlorinated biphenyls-induced oxidative stress by enhancing antioxidant enzymes activity in tomato. *Chemosphere* **2013**, *90*, 2645–2653. [CrossRef]
47. Mei, Y.; Chen, H.; Shen, W.; Huang, L. Hydrogen peroxide is involved in hydrogen sulfide-induced lateral root formation in tomato seedlings. *BMC Plant Biol.* **2017**, *17*, 162. [CrossRef]
48. Wang, Y.; Lv, P.; Kong, L.; Shen, W.; He, Q. Nanomaterial-mediated sustainable hydrogen supply induces lateral root formation via nitrate reductase-dependent nitric oxide. *Chem. Eng. J.* **2021**, *405*, 126905. [CrossRef]
49. Herschbach, C.; Pilch, B.; Tausz, M.; Rennenberg, H.; Grill, D. Metabolism of reduced and inorganic sulphur in pea cotyledons and distribution into developing seedlings. *New Phytol.* **2002**, *153*, 73–80. [CrossRef]
50. Chen, Y.; Lu, M.; Guo, D.; Zhai, Y.; Miao, D.; Yue, J.; Yuan, C.; Zhao, M.; An, D. Antifungal effect of magnolol and honokiol from *Magnolia officinalis* on *Alternaria alternata* causing tobacco brown spot. *Molecules* **2019**, *24*, 2140. [CrossRef]
51. Chaerani, R.; Groenwold, R.; Stam, P.; Voorrips, R.E. Assessment of early blight (*Alternaria solani*) resistance in tomato using a droplet inoculation method. *J. Gen. Plant Pathol.* **2007**, *73*, 96–103. [CrossRef]
52. Xue, Q.; Chen, Y.; Li, S.; Chen, L.; Ding, G.; Guo, D.; Guo, J. Evaluation of the strains of *Acinetobacter* and *Enterobacter* as potential biocontrol agents against Ralstonia wilt of tomato. *Biol. Control* **2009**, *48*, 252–258. [CrossRef]
53. Cakmak, I.; Marschner, H. Magnesium deficiency and high light intensity enhance activities of superoxide dismutase, ascorbate peroxidase, and glutathione reductase in bean leaves. *Plant Physiol.* **1992**, *98*, 1222–1227. [CrossRef] [PubMed]

54. Livak, K.J.; Schmittgen, T.D. Analysis of relative gene expression data using real-time quantitative PCR and the $2^{-\Delta\Delta C_T}$ method. *Methods* **2001**, *25*, 402–408. [CrossRef] [PubMed]
55. Qian, M.; Li, X.; Shen, Z. Adaptive copper tolerance in Elsholtzia haichowensis involves production of Cu-induced thiol peptides. *Plant Growth Regul.* **2005**, *47*, 65–73. [CrossRef]
56. Gong, B.; Li, X.; Bloszies, S.; Wen, D.; Sun, S.; Wei, M.; Li, Y.; Yang, F.; Shi, Q.; Wang, X. Sodic alkaline stress mitigation by interaction of nitric oxide and polyamines involves antioxidants and physiological strategies in Solanum lycopersicum. *Free Radic. Biol. Med.* **2014**, *71*, 36–48. [CrossRef]
57. Schreiber, U.; Schliwa, U.; Bilger, W. Continuous recording of photochemical and non-photochemical chlorophyll fluorescence quenching with a new type of modulation fluorometer. *Photosynth. Res.* **1986**, *10*, 51–62. [CrossRef]
58. Su, J.; Nie, Y.; Zhao, G.; Cheng, D.; Wang, R.; Chen, J.; Zhang, S.; Shen, W. Endogenous hydrogen gas delays petal senescence and extends the vase life of lisianthus cut flowers. *Postharvest Biol. Technol.* **2019**, *147*, 148–155. [CrossRef]
59. Gong, B.; Wen, D.; VandenLangenberg, K.; Wei, M.; Yang, F.; Shi, Q.; Wang, X. Comparative effects of NaCl and NaHCO3 stress on photosynthetic parameters, nutrient metabolism, and the antioxidant system in tomato leaves. *Sci. Hortic.* **2013**, *157*, 1–12. [CrossRef]
60. Bradford, M.M. A rapid and sensitive method for the quantitation of microgram quantities of protein utilizing the principle of protein-dye binding. *Anal. Biochem.* **1976**, *72*, 248–254. [CrossRef]

Article

Integrated Metabolomic and Transcriptomic Analyses to Understand the Effects of Hydrogen Water on the Roots of *Ficus hirta Vahl*

Jiqing Zeng [1],* and Hui Yu [2,3]

1   Key Laboratory of South China Agricultural Plant Molecular Analysis, Genetic Improvement Guangdong Provincial Key Laboratory of Applied Botany, South China Botanical Garden, Chinese Academy of Sciences, Guangzhou 510650, China
2   Key Laboratory of Plant Resource Conservation and Sustainable Utilization, Guangdong Provincial Key Laboratory of Applied Botany, South China Botanical Garden, Chinese Academy of Sciences, Guangzhou 510650, China; yuhui@scbg.ac.cn
3   Southern Marine Science and Engineering Guangdong Laboratory (Guangzhou), Guangzhou 511458, China
*   Correspondence: zengjq@scib.ac.cn

Abstract: Wuzhimaotao (*Ficus hirta Vahl*) is an important medicinal and edible plant in China. The extract from the roots of *Ficus hirta Vahl* contains phenylpropanoid compounds, such as coumarins and flavonoids, which are the main active components of this Chinese herbal medicine. In this study, we analyzed the transcriptomic and metabolomic data of the hydrogen-water-treated roots of *Ficus hirta Vahl* and a control group. The results showed that many genes and metabolites were regulated in the roots of *Ficus hirta Vahl* that were treated with hydrogen water. Compared with the control group, 173 genes were downregulated and 138 genes were upregulated in the hydrogen-rich water treatment group. Differential metabolite analysis through LC-MS showed that 168 and 109 metabolites had significant differences in positive and negative ion mode, respectively. In the upregulated metabolites, the main active components of Wuzhimaotao, such as the phenylpropane compounds naringin, bergaptol, hesperidin, and benzofuran, were found. Integrated transcriptomic and metabolomic data analysis showed that four and one of the most relevant pathways were over enriched in positive and negative ion mode, respectively. In the relationship between metabolites and DEGs, phenylpropanoid biosynthesis and metabolism play an important role. This indicates that phenylpropanoid biosynthesis and metabolism may be the main metabolic pathways regulated by hydrogen water. Our transcriptome analysis showed that most of the DEGs with $|\log2FC| \geq 1$ are transcription factor genes, and most of them are related to plant hormone signal transduction, stress resistance, and secondary metabolism, mainly phenylpropanoid biosynthesis and metabolism. This study provides important evidence and clues for revealing the botanical effect mechanism of hydrogen and a theoretical basis for the application of hydrogen agriculture in the cultivation of Chinese herbal medicine.

Keywords: Wuzhimaotao (*Ficus hirta Vahl*); hydrogen; transcription factors; secondary metabolism; phytohormones signaling pathways; phenylpropanoid biosynthesis and metabolism; Chinese herbal medicine

Citation: Zeng, J.; Yu, H. Integrated Metabolomic and Transcriptomic Analyses to Understand the Effects of Hydrogen Water on the Roots of *Ficus hirta Vahl*. *Plants* **2022**, *11*, 602. https://doi.org/10.3390/plants11050602

Academic Editor: John Hancock

Received: 29 December 2021
Accepted: 9 February 2022
Published: 24 February 2022

1. Introduction

Wuzhimaotao (*Ficus hirta Vahl*) is a common Chinese herbal medicine in the Lingnan area of China. It is a folk homologous plant of food and medicine, which is called Guangdong ginseng. It is used by Hakka people in southern China to treat diseases such as spleen deficiency, tuberculosis, weakness, rheumatism, night sweats, and agalactism [1–3]. Wuzhimaotao uses the roots of *Ficus hirta Vahl* as medicine. It is mainly distributed in Guangdong, Guangxi, Jiangxi, Fujian, Yunnan, Hong Kong of China, and Southeast Asian countries. The isolated and identified compounds in Wuzhimaotao mainly include coumarins, flavonoids, and volatile oil. Modern pharmacological studies show that these compounds have antioxidant, anti-inflammatory, antibacterial, antiviral, and antitumor effects [4].

In fact, the main pharmacodynamic components of Chinese herbal medicine are usually plant secondary metabolites. Studies shows that stress plays an important role in the formation of genuine medicinal materials [5]. When plants are infected by biological or abiotic factors, they can synthesize and accumulate a series of low relative molecular weight compounds with disease resistance through the expression of resistance genes in vivo to resist the invasion of pathogens. These substances are commonly known as phytoalexins. Phytoalexins are usually secondary metabolites, which are a large class of small molecular compounds produced by plants in the process of secondary metabolism [6]. Common secondary metabolites include alkaloids, flavonoids, terpenoids, anthraquinones, coumarins, and lignins. Stress conditions can promote the accumulation of plant secondary metabolites, and the accumulation of secondary metabolites can improve the stress resistance of plants. Therefore, the measures to improve plant stress resistance are expected to promote the accumulation of plant secondary metabolites in an effort to improve the genuine quality of traditional Chinese medicine [5]. The active components (coumarins, flavonoids, and volatile oil) of Wuzhimaotao are common plant secondary metabolites. The quality of the traditional Chinese medicine Wuzhimaotao and the content of its active components are expected to improve as stress conditions or other methods are used to improve its stress resistance.

We found that hydrogen is involved in plant stress responses by regulating plant hormone signal transduction [7]. Hydrogen water treatment can improve the disease resistance of plants and the salt and drought resistance of rice seedlings. Stress treatment can also improve the ability of rice to release hydrogen [7]. Hydrogen water treatment can reduce the harm of heavy metal stress to plants [8–10]. We also found that hydrogen water can promote the growth of mung bean and rice roots and seedlings [7]. Therefore, we put forward the concept of hydrogen agriculture, i.e., using hydrogen water and other kinds of hydrogen fertilizer to promote the growth of crops and improve their stress resistance, in order to reduce the use of pesticides and chemical fertilizers, protect the ecological environment, and ensure food safety [11]. In addition, hydrogen fertilizers could be applied to the cultivation of traditional Chinese medicine to improve the content of effective components, yield, stress resistance, and, therefore, the quality of genuine medicinal materials.

In order to determine whether hydrogen molecules can increase the content of plant secondary metabolites and the molecular mechanism of hydrogen regulating the synthesis of secondary metabolites, the metabolome and transcriptome of Wuzhimaotao were analyzed.

2. Results

2.1. Metabolomic Changes Associated with Hydrogen-Water-Treated Roots of Ficus hirta Vahl

To investigate the metabolic pathways that were perturbed in the roots of *Ficus hirta Vahl* by the hydrogen water, we performed an integrated analysis of metabolomics and transcriptomics. Using liquid chromatography mass-spectrometry (LC-MS) for untargeted metabolomic profiles, a total of 12 samples was detected, including 6 samples of hydrogen-water-treated roots and 6 controls.

In order to further understand the differences between hydrogen-water-treated and control groups, we use Principal Component Analysis (PCA) for metabolite composition analysis, and we performed partial least squares discriminant analysis (PLS-DA) to sharpen the separation between groups of observations by rotating PCA components such that a maximum separation among classes was obtained and variables carrying the class separating information were located. Data of 1018 metabolites under the positive analysis ion mode and of 769 metabolites under the negative analysis ion mode were used in PLS-DA. Figure 1 shows that the score was 27.56% in ESI+ mode and 22.31% in ESI− mode. The hydrogen water treatment group could be clearly separated from the control group in the *x*-axis direction according to the PLS-DA model. The model quality was determined by parameters R2Y and Q2Y; R2Y = 0.99, Q2Y = 0.75 (Figure 1). The PLSDA score plot showed a clear separation between hydrogen-water-treated samples and control samples under the ESI− mode, in which R2Y and Q2Y values were greater than 0.5 (Figure 1). Therefore, the PLS-DA model with the first two components is sufficient to explain the difference between the hydrogen-water-treated samples and the control samples.

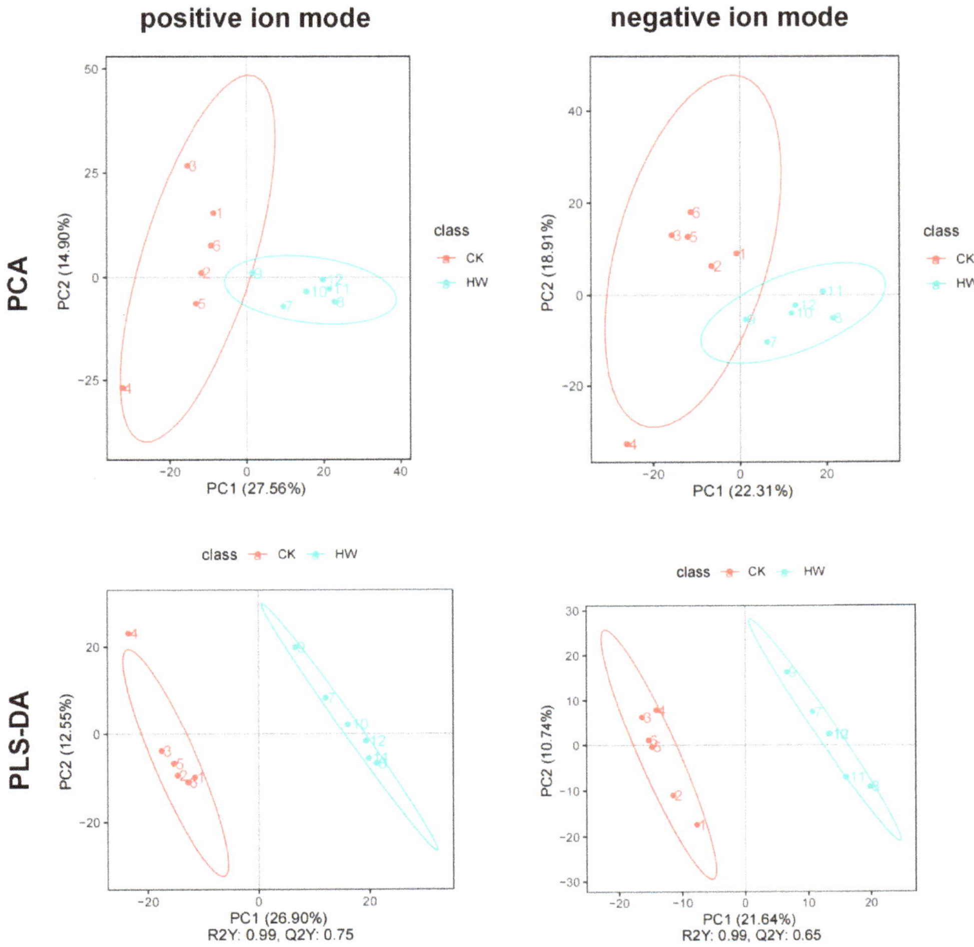

Figure 1. PCA and partial least squares discriminate analysis (PLS-DA) score chart of metabolite profiling data in positive ion mode and negative ion mode.

Significantly differentially expressed metabolites were identified according to the following criteria: PLS-DA VIP (variable importance in the projection) > 1, fold change > 2.0, and *p*-value < 0.05. A total of 1018 differentially expressed metabolites in hydrogen-water-treated samples compared to the control were identified in positive ion mode. A total of 168 significantly different metabolites were identified; 84 metabolites were upregulated while the other 84 metabolites were downregulated in hydrogen-water-treated samples. A total of 769 metabolites were identified in the negative ion mode. Compared with the control group, there were 109 significantly differentially expressed metabolites in the hydrogen water treatment group, including 64 upregulated metabolites and 45 downregulated metabolites (Table S1). Figure 2A,C show that all of these metabolites were included in the multivariate analysis. These results showed that hydrogen water treatment led to significant changes in root metabolism of *Ficus hirta Vahl*.

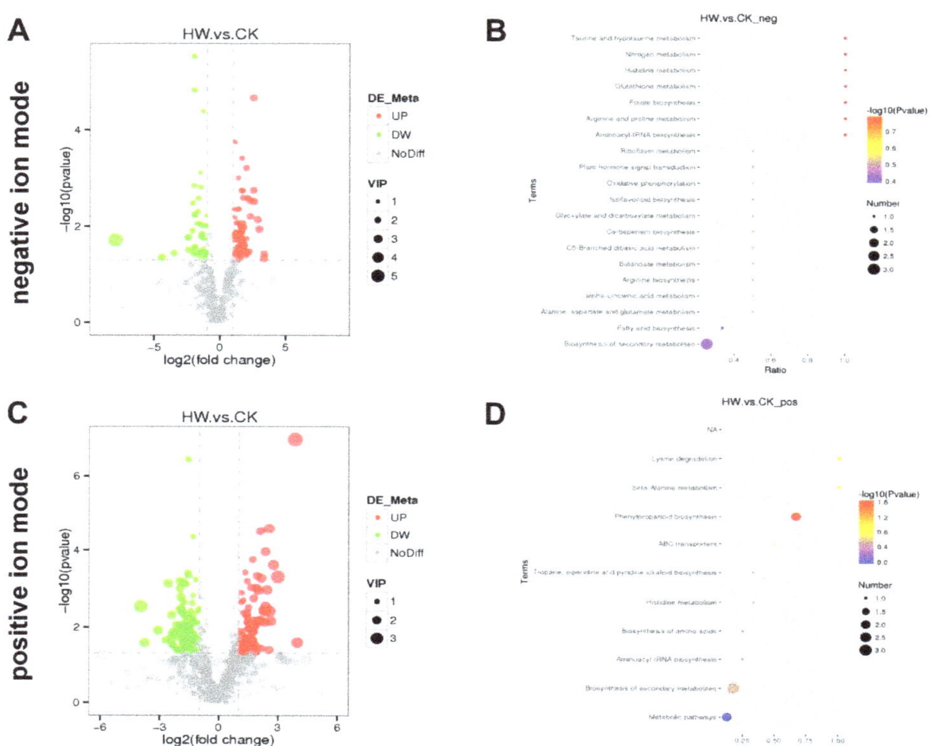

Figure 2. Metabolomics analysis of the hydrogen-water-treated group (HW) and the control group (CK). Volcano figure under negative ion mode (**A**) and positive ion mode (**C**); the green points represent the downregulated metabolites and the red points represent upregulating in the hydrogen water treated group. For the metabolite enrichment analysis under negative ion mode (**B**) and positive ion mode (**D**), the abscissa represents the ratio of the number of differential metabolites in the corresponding path to the total number of identified metabolites and the ordinate represents the impact of the path. The statistical significance is expressed in color gradient, and the size of the circle is directly proportional to the number of metabolites.

2.2. Pathway Enrichment Analysis

In order to investigate the effect on related pathways regulated in roots of *Ficus hirta Vahl* by hydrogen water, pathway enrichment analysis was performed. Ten and twenty-three related metabolic pathways were identified in ES+ and ESI− mode, respectively. The results showed that the significantly different metabolites mainly associated with phenylpropanoid biosynthesis ($p = 0.022$) in ESI+ mode (Figure 2D). However, under the ESI− mode, lysine degradation ($p = 0.159$) and beta-alanine metabolism ($p = 0.159$) are close to significant (Figure 2B). The KEGG map of phenylpropanoid biosynthesis was shown in Figure S1.

2.3. Differentially Expressed Genes (DEGs) between Hydrogen Water Treatment Groups and Control Groups

To understand the molecular basis of the metabolic pathway in roots of *Ficus hirta Vahl*, transcriptomes were analyzed to identify differentially expressed genes. RNA-seq data were from the hydrogen water treatment group and the control group, with three biological replicates respectively.We identified differentially expressed genes with the DESeq package by using the following criteria: fold change > 2 and adjusted *p*-value < 0.05. The obvious expressed unigenes with red and blue colors in two groups were clearly reflected in the Volcano plot. The results showed that 311 differentially expressed genes (DEG) were identified in roots of *Ficus hirta Vahl*, of which 138 were upregulated and 173 were downregulated (Figure 3A). Table S2 lists the upregulated and downregulated genes with significant differences between the two groups. The specificity and coexpression of mRNAs between the two groups are shown in the Venn diagram (Figure 3B). Transcripts clusters with significantly different expression patterns between the treatment group and the control group are highlighted on the heat map (Figure 3C). These results showed that hydrogen water treatment had a significant effect on transcription.

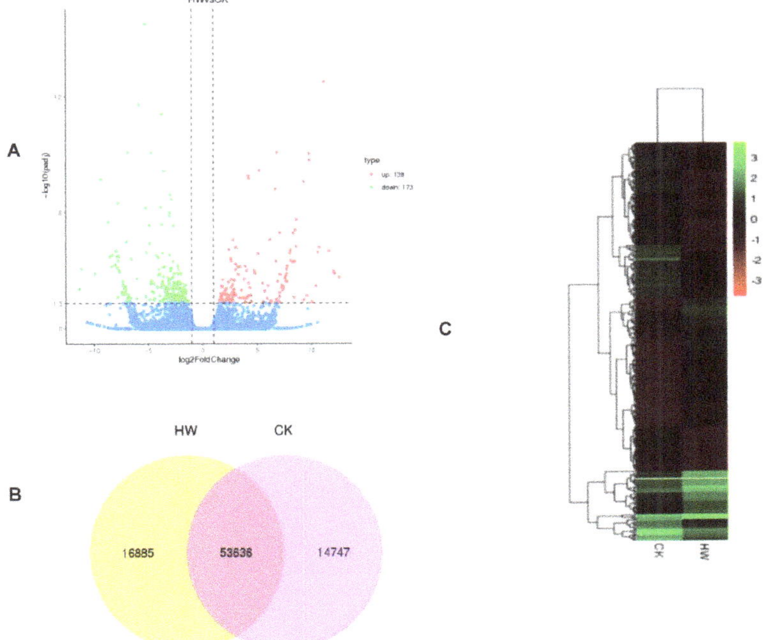

Figure 3. Cluster analysis of differentially expressed genes. (**A**) Volcano map of differentially expressed mRNAs. (**B**) Venn plot of the number of single genes between the two treatment groups. (**C**) Heat map of differentially expressed genes between hydrogen water treatment group and control group.

In order to further characterize DEGs, the KEGG pathway database was used for pathway analysis. The results showed that the transcripts regulated by hydrogen water in *Ficus hirta Vahl* roots could be mapped to signal pathways, such as carotenoid biosynthesis, phenylpropanoid biosynthesis, plant-pathogen interaction, carbon fixation in photosynthetic organisms, and starch and sucrose metabolism (Figures 4A,B and S2). The gene ontology (GO) analysis of differentially expressed genes was classified based on GO annotation terms, and the gene expression pattern of the effect of hydrogen water on *Ficus hirta Vahl* roots was obtained. Figure 4C shows the enriched GO terms interaction network in biological processes, cellular components and molecular functions, and evaluates the relationship between differentially expressed genes.

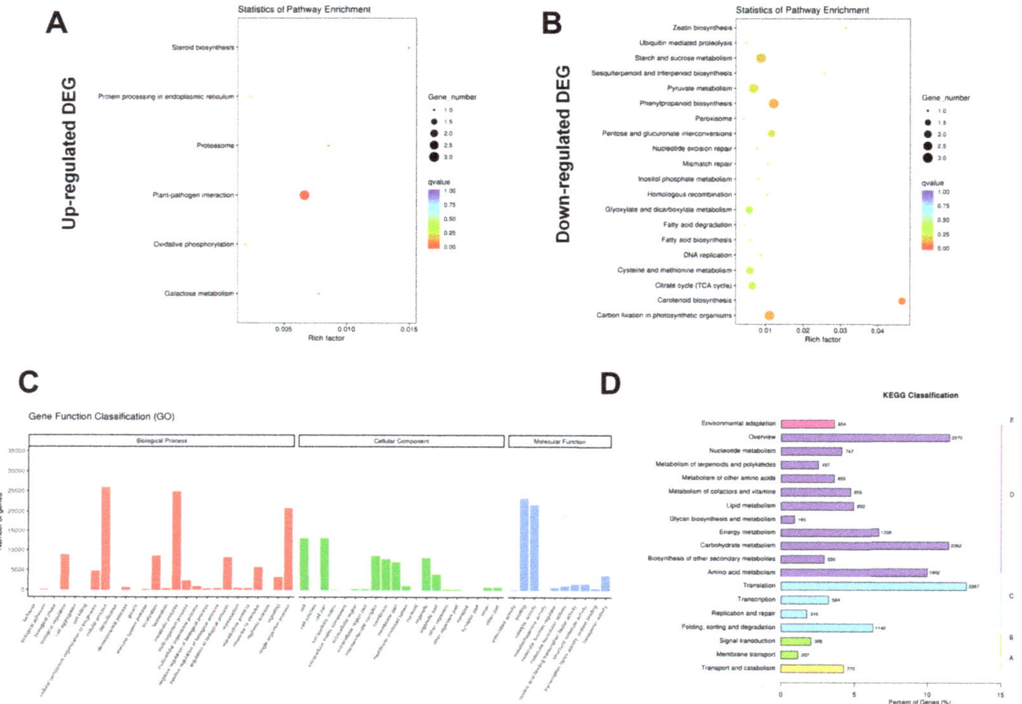

Figure 4. Functional analysis of DEGs between hydrogen water treatment group (HW) and control group (CK) based on Gene Ontology and KEGG pathway. (**A**) Scatter diagram of upregulated DEGs and enriched KEGG pathway; (**B**) scatter diagram of down regulated DEGs and enriched KEGG pathway; (**C**) enriched interaction network of GO terms in biological processes, cellular components, and molecular functions; (**D**) KEGG classification.

2.4. Transcription Factors Related to Plant Hormone Signal Transduction, Stress Resistance, and Secondary Metabolite Biosynthesis in Roots of Ficus hirta Vahl

Transcription factors participate in plant hormone signal transduction, stress resistance, and secondary metabolite biosynthesis by regulating the gene expression in plants. In our data, we found 625 transcription factors with different expression levels using iTAK software (http://itak.feilab.net/cgi-bin/itak/index.cgi/, accessed on 28 December 2021), of which, 168 transcription factors with significantly differentially expressed levels were related to phenylpropanoid biosynthesis and defense signaling (Table S3). Among these transcription factors, the most abundant were MYBs (41) and bHLHs (27), followed by AP2/ERFs (26), bZIPs (20), NACs (16), WRKYs (14), AUX/IAAs (12), MADs (10), and HSFs (2) (Figure 5). These transcription factors contribute to plant hormone signal transduction, stress resistance, and secondary metabolite biosynthesis in *Ficus hirta Vahl* roots.

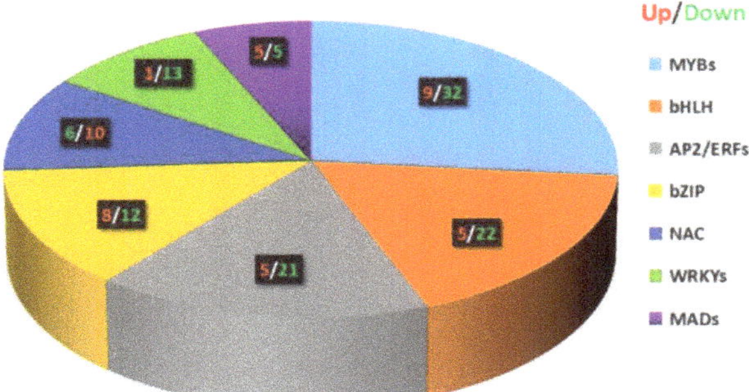

Figure 5. Transcription factors of phenylpropanoid biosynthesis in the roots of *Ficus hirta Vahl.*

2.5. Integrated Analysis of the Mechanism of Hydrogen-Water-Treated Ficus hirta Vahl Root from Metabolomic and Transcriptomic Data

We further integrated metabolomic and transcriptomic data at the pathway level to explore the potential relationship between DEGs and metabolites. The results show that there are four enrichment pathways in ESI− mode, namely plant hormone signal transduction, oxidative phosphorylation, glyoxylic acid and dicarboxylic acid metabolism, and fatty acid biosynthesis (Figure 6), and only one enriched pathway under ESI+ mode, namely phenylpropanoid biosynthesis. Phenylpropanoid biosynthesis plays an important role in the potential relationship between DEGs and metabolites. These results suggest that phenylpropanoid biosynthesis might be the main metabolic pathway regulated by hydrogen water in *Ficus hirta Vahl* root. Figure 7 shows the heat map of comprehensive analysis between differential metabolite expression patterns and transcriptomics in ESI− mode and ESI+ mode.

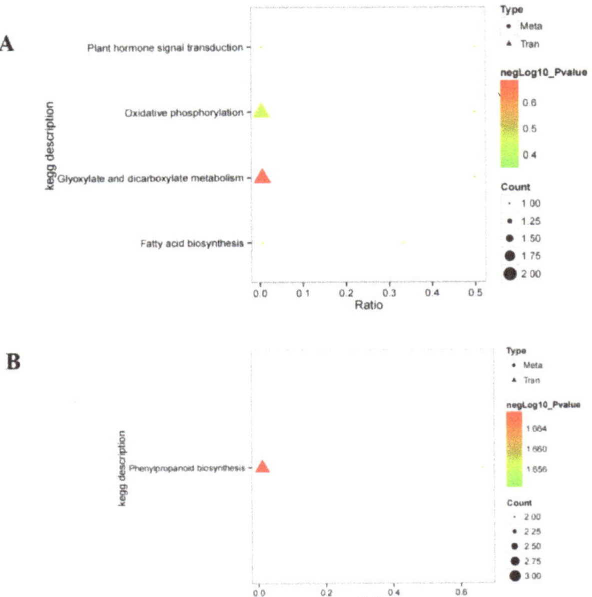

Figure 6. Integrated altered metabolic pathways in hydrogen-water-treated *Ficus hirta Vahl* roots. (**A**) negative ion mode; (**B**) positive ion mode.

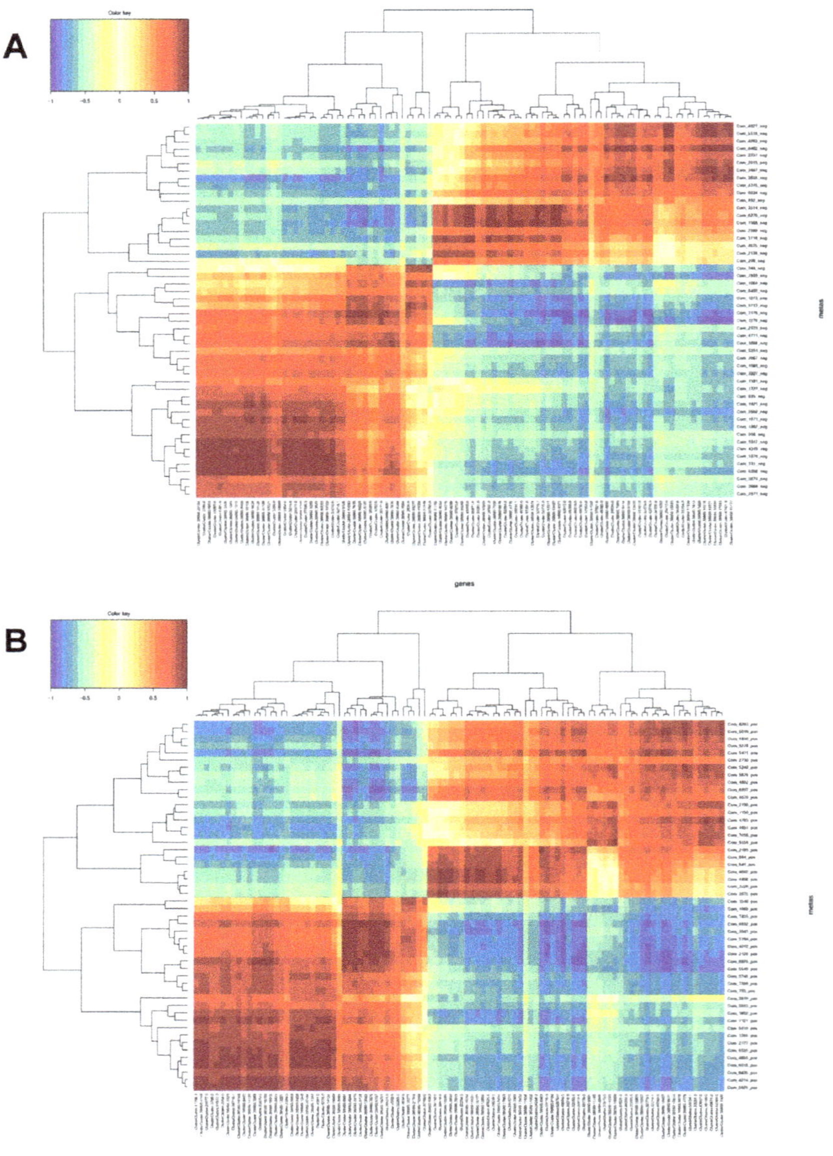

Figure 7. Integrative analysis based on metabolomics and transcriptomics data. (**A**) Negative ion mode. (**B**) Positive ion mode. The horizontal axis represents the differentially expressed gene ID number, and the vertical axis represents differential metabolite ID number. The blue depth indicates the intensity of the negative correlation. The red depth indicates the intensity of the positive correlation.

3. Discussion

Although the biological effects of hydrogen have long been noticed in botany and medicine [12–14], they attracted extensive attention only after 2007 [15]. Presently, it is understood that the mechanism of the hydrogen biological effect is that hydrogen has both antioxidant effects and signal molecule effects [16,17]. Our previous study found that hydrogen affects plant growth, development, and stress adaptation by participating in the regulation of plant hormone signaling pathways [7,18]. Because the improvement of plant stress resistance is often manifested in the accumulation of secondary metabolites [19–21], cultivation using hydrogen water may improve the accumulation of secondary metabolites, which may be conducive to improving the quality of traditional Chinese medicine.

Wuzhimaotao is a famous traditional herbal medicine with homology of medicine and food in southern China that has been used for centuries. It is the dry root of *Ficus hirta*, a mulberry plant. The antioxidant, anti-inflammatory, antibacterial, antiviral, and antitumor effects of Wuzhimaotao have attracted more and more attention. It was reported that the main active components of Wuzhimaotao include coumarins, such as isopsoralen lactone, bergamot lactone, and psoralen, as well as flavonoids, such as apigenin and hesperidin [4]. It was reported that 70 compounds were isolated in Wuzhimaotao, of which 30 phenylpropanoid compounds were the most abundant [22]. The phenylpropane metabolic pathway in plants is an important pathway of secondary metabolism. All substances containing a phenylpropane skeleton are direct or indirect products of this pathway [23]. The phenylpropanoid biosynthetic pathway is activated under stress conditions, resulting in accumulation of various phenolic compounds which, among other roles, have the potential to scavenge harmful reactive oxygen species (ROS) [24–26]. In recent years, metabolomics integrated with transcriptomics has been widely used to investigate the biosynthesis of metabolites in plants [27,28]. In this study, hydrogen water was used to treat the famous Chinese southern medicine Wuzhimaotao, and then the metabolome and transcriptome were analyzed. Metabolomic analysis showed that there were a total of 277 significantly different metabolites between the hydrogen-water-treated group and the control group, including 148 upregulated and 129 downregulated metabolites. Among the upregulated metabolites, naringin (ID: Com_788_neg; fold change: 3.05), bergaptol (ID: Com_1406_neg; fold change: 2.28), hesperidin (ID: Com_3438_pos; fold change: 2.25), and a benzofuran (ID: Com_1053_neg; fold change: 2.19) were found (Table S1), all of which are phenylpropane compounds. As their congeners and derivatives, narigenin, bergapten, hesperidin, and benzofuran compounds of were isolated and identified as the active ingredients in Wuzhimaotao. Psoralen, a benzofuran derivative, is one of the most important active components of Wuzhimaotao. Recently, many benzofuran compounds were isolated from Wuzhimaotao, such as 3-[6-(5-O-β-D-glucopyranosyl) benzofuranyl] methyl propionate, (E)-3-[5-(6-hydroxy) benzofuranyl] propenoic acid, (E)-3-[5-(6-methoxy) benzofuranyl] propenoic acid, (Z)-3-[5-(6-O-β-D-glucopyranosyl) benzofuranyl] methyl propenoate, and S-3-[2,3-dihydro-6-hydroxy-2-(1-hydroxy-1-methylethyl)-5-benzofuranyl] methyl propionate [22]. Bergapten, psoralen, and other benzofurans belong to coumarins, and naringin and hesperidin belong to flavonoids. Coumarins and flavonoids are formed through the phenylpropanoid biosynthesis pathway. The results of the pathway enrichment analysis show that the significantly different metabolites were mainly associated with phenylpropanoid biosynthesis. The results show that hydrogen water treatment promoted the accumulation of active components in Wuzhimaotao, many of which were produced by phenylpropane synthesis. Therefore, planting Wuzhimaotao with hydrogen water may help to improve the content of the active components of Wuzhimaotao and improve the quality of traditional Chinese medicine.

Transcriptome analysis showed that there were 311 significantly different DEGs between the hydrogen water treatment group and the control group, including 138 upregulated and 173 downregulated unigenes. KEGG pathway analysis found that transcripts regulated in the hydrogen-water-treated *Ficus hirta Vahl* roots could be mapped to signaling pathways, such as carotenoid biosynthesis, phenylpropanoid biosynthesis, and plant-pathogen interaction. The results of combined analysis of the metabolome and transcriptome showed that there were five enriched pathways: phenylpropanoid biosynthesis, plant hormone signal transduction, oxidative phosphorylation, glyoxylate and dicarboxylate metabolism, and fatty acid biosynthesis. The results show that hydrogen water treatment affected the phenylpropanoid biosynthesis and plant hormone signal transduction in *Ficus hirta Vahl* roots. Figure 4B shows that the DEGs of the phenylpropane synthesis pathway are downregulated, which seems to be inconsistent with the upregulation of coumarin and flavonoid metabolites shown by metabolome analysis. Comparing the KEGG enrichment obtained from metabolome and transcriptome analysis, we found that the phenylpropane synthesis pathway of the metabolome (Figure S1) is different from that of the transcriptome (Figure S2). Figure S2 shows that the metabolic pathways of coumarins and flavonoids are not downregulated. Therefore, there is no contradiction between the downregulation of the phenylpropane synthesis pathway shown by transcriptome analysis and the upregulation of coumarin and flavonoid metabolites shown by metabolome analysis. In addition, the integrated analysis of the metabolome and transcriptome showed that there was a significant positive or negative correlation between differential metabolites and differentially expressed genes (Figure 7). Therefore, the downregulation of genes related to the phenylpropane metabolic pathway is reflected in the downregulation of genes shown in Figure S2 and may also show that some negatively regulated transcription factor genes are downregulated.

The biosynthesis of secondary metabolites such as phenylpropanes is mainly regulated by transcription factors at the transcriptional level [29]. Recently, it has been found that six TF families are closely related to defense signaling. These six TF families are: AP2/ERF (APETALA2/ethylene responsive factor), bHLH (basic helix-loop-helix), MYB (myeloblastosis related), NAC (no apical meristem (NAM), Arabidopsis transcription activation factor (ATAF1/2), cup-shaped cotyledon (CUC2)), WRKY, and bZIP (basic leucine zipper) [30,31]. Interestingly, our transcriptome analysis showed that most of the DEGs with | log2FC | $\geq$ 1 are transcription factor genes. Among them, 41 MYB, 27 bHLH, 26 AP2/ERF, 20 bZIP, 16 NAC, 14 WRKY, 12 AUX/IAA, 10 MAD, and 12 HSF transcription factor genes were found. Clearly, these transcription factors are related to plant hormone signal transduction, stress resistance, and secondary metabolite synthesis. Since most of the transcription factors related to the phenylpropane metabolic pathway shown in Figure 6 are downregulated, we speculate that the downregulation of these transcription factors leads to the downregulation of the DEGs shown in Figure S2. It is also possible that some transcription factors play a negative regulation role and the down-regulation of transcription factors promote the coumarin and flavonoid metabolic pathways in the phenylpropane metabolic pathway. In addition, we did not rule out the possibility that the downregulation of other pathways of the phenylpropane metabolic pathway can promote the metabolism of coumarins and flavonoids.

Our previous study found that stress conditions can promote the release of hydrogen in plants [7]. Hydrogen can regulate the expression of transcription factor genes related to stress and plant hormone signal transduction, which suggests that hydrogen may be the secondary messenger under stress conditions. Since plant stress resistance and the accumulation of secondary metabolites are related to the regulation of plant hormones [7], these results suggest that hydrogen may act as a signal molecule to regulate the expression of transcription factor genes related to plant hormone and defense signal transduction, which is consistent with our previous research conclusions.

4. Materials and Methods

4.1. Plant Materials and Treatments

Twelve Wuzhimaotao (*Ficus hirta Vahl*) plants growing in South China Botanical Garden were selected and evenly divided into two groups, the treatment group and the control group. Hydrogen water was obtained following the method of Li et al. [32]. Purified hydrogen gas (99.99%, *v/v*) (Kedi Gas Chemical Co., Ltd., Foshan, China) was bubbled into 5 L of pure water at a rate of 200 mL min^{-1} for 3 h to obtain the saturated hydrogen water. The hydrogen concentration of hydrogen water was determined using a hydrogen portable meter (Trustlex Co., Ltd., ENH-1000, Yokohama, Japan). The treatment group was watered with hydrogen water (0.8 ppm) once a week for a total of 3 times, while the control group was watered with pure water. On the 15th day, the roots of *Ficus hirta Vahl* were collected, frozen with liquid nitrogen and stored at $-80\ ^\circ$C before metabolite extraction and RNA-Seq.

4.2. Metabolite Extraction and Analysis

Each sample tissue (100 mg) was grounded with liquid nitrogen, and the homogenate was resuspended with precooled 80% methanol and 0.1% formic acid by well vortexing. The sample was placed on ice for 5 min and then centrifuged at 15,000 rpm at 4 $^\circ$C for 5 min. Part of the supernatant was diluted with methanol to a final concentration of 60% with LC-MS grade water.

The samples were subsequently transferred to a fresh Eppendorf tube with a 0.22 µm filter and then were centrifuged at 15,000× g at 4 $^\circ$C for 10 min. Finally, the filtrate was injected into the LC-MS/MS system analysis. Liquid sample (100 µL) and prechilled methanol (400 µL) were mixed by well vortexing.

The Vanquish UHPLC system (Thermo Fisher, MA, USA) and Orbitrap Q Exactive series mass spectrometer (Thermo Fisher, MA, USA) were used for LC-MS/MS analysis. The flow rate of sample injected into a Hyperil Gold column (100 × 2.1 mm, 1.9 µm) was 0.2 mL/min, and the linear gradient retention time was 16 min. Eluent A (0.1% FA in water) and eluent B (Methanol) were used for positive mode. Eluent A (5 mM ammonium acetate, pH 9.0) and eluent B (Methanol) were used for negative polarity mode. The setting of solvent gradient was: 2% B, 1.5 min; 2–100% B, 12.0 min; 100% B, 14.0 min; 100–2% B, 14.1 min; and 2% B, 16 min. Under the positive and negative polarity mode, the Q Exactive mass spectrometer was run with a spray voltage of 3.2 kV, the capillary temperature at 320 $^\circ$C, a sheath gas velocity of 35arb, and an auxiliary gas velocity of 10arb.

For peak alignment, pickup, and quantification of each metabolite, Compound Discoverer 3.0 (CD3.0, Thermo Fisher) was used to process the raw data file generated by UHPLC-MS/MS. The main parameters were as follows: retention time tolerance was 0.2 min; the actual mass tolerance was 5 ppm; signal strength tolerance was 30%; the signal-to-noise ratio was 3 and minimum intensity was 100,000. Then, the peak intensity was normalized to the total spectral intensity. Normalized data can be used to predict molecular formulas based on molecular ion peaks, additive ions, and fragment ions. Then, in order to obtain accurate qualitative and relative quantitative results, the peaks were matched with the mzCloud (https://www.mzcloud.org/, accessed on 23 May 2020) and ChemSpider (http://www.chemspider.com/, accessed on 23 May 2020) databases. Statistical analyses were performed using the statistical software R (R version R-3.4.3), Python (Python 2.7.6 version), and CentOS (CentOS release 6.6). When the data did not conform to the normal distribution, the area normalization method was used for normal transformation.

These metabolites were annotated using the HMDB database (http://www.hmdb.ca/, accessed on 23 May 2020), KEGG database (http://www.genome.jp/kegg/, accessed on 23 May 2020), and Lipidmaps database (http://www.lipidmaps.org/, accessed on 23 May 2020). Principal component analysis (PCA) and partial least squares discriminant analysis (PLS-DA) were performed in metaX. Statistical significance (p-value) was calculated by univariate analysis (t-test). The identification criteria of differential metabolites were VIP > 1, p-value < 0.05, and fold change ≥ 2 or FC ≤ 0.5. The metabolites of interest were filtered according to the log2 (FC) and-log10 (p-value) of metabolites to obtain the volcano map.

The data were normalized using the Z score of the intensity region of differential metabolites, and the clustering heat map was drawn using the phatmap software package in R language. Cor() in R language (method = pearson) was used to analyze the correlation between different metabolites. We used cor.mtest() in R language to calculate the statistically significant correlation between different metabolites. The p-value with statistical significance was <0.05, and the correlation diagram was drawn with the corrplot software package in R language. We used the KEGG database to study the function and metabolic pathways of these metabolites. The criteria for metabolic pathway enrichment of differential metabolites are as follows. The metabolic pathway enrichment of differential metabolites were performed, when ratio were satisfied by $x/n > y/N$, metabolic pathway were considered as enrichment, when p-value of metabolic pathway < 0.05, metabolic pathway were considered as statistically significant enrichment.

4.3. Total RNA Extraction, RNA Quantification and Qualification

Frozen samples were ground in liquid nitrogen. Total RNA was extracted using a TRIzol reagent. We used 1% agarose gel electrophoresis to monitor RNA degradation and pollution. The NanoPhotometer® spectrophotometer (IMPLEN, Westlake Village, CA, USA) was used to check RNA purity. The RNA Nano 6000 Assay Kit of the Agilent Bioanalyzer 2100 system (Agilent Technologies, Santa Clara, CA, USA) was used to assess RNA integrity.

4.4. Library Preparation for Transcriptome Sequencing

The total amount of RNA in each sample used as input material for RNA sample preparation was 1.5 µg. Following the manufacture's recommendations and adding index codes to attribute sequences to each sample, a NEBNext® Ultra™ RNA Library Prep Kit for Illumina® (NEB, Ipswich, MA, USA) was used to generate sequencing libraries. In short, mRNA was purified from total RNA by poly-Toligo-attached magnetic beads.

Under elevated temperatures, divalent cations were used for cleavage in the NEBNext First Strand Synthesis Reaction Buffer (5X). The first strand of cDNA was synthesized by random hexamer primer and M-MuLV Reverse Transcriptase (RNase H-). The second strand of cDNA was then synthesized using DNA Polymerase I and RNase H. The remaining overhangs were transformed into blunt ends by exonuclease/polymerase activities. After adenylation of 3′ ends of DNA fragments, NEBNext Adaptor with hairpin loop structure were ligated to prepare for hybridization.Library fragments were purified using the AMPure XP system (Beckman Coulter, Beverly, USA) to preferentially select cDNA fragments with a length of 250~300 bp. Then 3 µL of USER Enzyme (NEB, USA) was used with size-selected, adaptor-ligated cDNA at 37 °C for 15 min followed by 5 min at 95 °C before PCR. PCR was then performed with Phusion High-Fidelity DNA polymerase, Universal PCR primers, and Index (X) Primer. Finally, the PCR products were purified (AMPure XP system) on the Agilent Bioanalyzer 2100 system and the library quality was evaluated.

4.5. Clustering and Sequencing (Novogene Experimental Department)

A TruSeq PE Cluster Kit v3-cBot-HS (Illumia) was used to cluster the index-coded samples on a cBot Cluster Generation System according to the manufacturer's instructions. The prepared libraries were then sequenced on the Illumina Hiseq platform, and paired end reads were generated.

4.6. Data Analysis

Quality control: Raw data (raw reads) in fastq format were processed through the internal Perl script. In this process, clean data (clean reads) were obtained by deleting the reads including adapter, poly-N, and low-quality reads from the raw data. In addition, Q20, Q30, GC content, and sequence repeat level of clean data needed to be calculated. All the downstream analyses were based on clean data with high quality.

Transcriptome assembly: The left files (read1 files) in all libraries and samples were merged into a large left.fq file, and the right files (read2 files) were merged into a large right.fq file. Transcriptome assembly was based on the left.fq and right.fq, and was completed using Trinity [33]; min_kmer_cov was set to 2 by default, and all other parameters were also set by default.

Gene function annotation: Gene functions were annotated based on Nr (NCBI nonredundant protein sequences), Nt (NCBI nonredundant nucleotide sequences), KOG/COG

(Clusters of Orthologous Groups of proteins), Pfam (Protein family), Swiss-Prot, KO (KEGG Ortholog database), and GO (Gene Ontology) databases.

Differential expression analysis: The differential expression of two groups of samples with biological duplication was analyzed by DESeq R package (1.10.1). DESeq provides statistical routines for determining differential expression in digital gene expression data using a model based on the negative binomial distribution. In order to control the error detection rate, the resulting p value was adjusted using the method of Benjamin and Hochberg. Genes found by DESeq were identified as differentially expressed genes if their adjusted p value was <0.05. The threshold of significant differential expression was set as q value < 0.005 and |log2 (foldchange)| > 1.

GO enrichment analysis: GO enrichment analysis of the differentially expressed genes (DEGs) was realized by the GOseq R packages based on Wallenius' noncentral hypergeometric distribution [34].

KEGG pathway enrichment analysis: KEGG [35] is a database resource (http://www.genome.jp/kegg/, accessed on 28 December 2021) used to understand the advanced functions and utilities of biological systems, such as cells, organisms, and ecosystems, from molecular level information, especially large-scale molecular data sets generated by genome sequencing and other high-throughput experimental technologies. KOBAS [36] software was used to test the statistical enrichment of differentially expressed genes in the KEGG pathway.

5. Conclusions

In order to study the changes of overall metabolite spectrum related to the effects of hydrogen water on *Ficus hirta Vahl* roots, metabolomics and transcriptomics were analyzed by LC-ESI-MS/MS and RNA-seq techniques. Our study demonstrated that hydrogen water led to significant metabolic alterations in roots of *Ficus hirta Vahl*. The significantly upregulated metabolites included narigenin, bergapten, hesperidin, and benzofuran compounds. These active components of Wuzhimaotao are flavonoids and coumarins, which are produced through phenylpropane synthesis. By integrating transcriptomic data obtained by RNA-seq, we indicated that phenylpropanoid biosynthesis and metabolism may be the major metabolic pathways disturbed by hydrogen water. Additionally, we revealed that molecular hydrogen may regulate the expression of transcription factor genes related to plant hormone signal transduction, stress resistance, and secondary metabolite synthesis. This study provides important clues and evidence for understanding the botanical effect mechanism of hydrogen and a theoretical basis for the application of hydrogen agriculture in the cultivation of Chinese herbal medicine.

Supplementary Materials: The following are available online at https://www.mdpi.com/article/10.3390/plants11050602/s1, Figure S1: KEGG map of phenypropanoid biosynthesis pathway from metabolome analysis, Figure S2: KEGG map of phenypropanoid biosynthesis pathway from transcriptome analysis, Table S1: Metabolites differentially expressed in hydrogen rich water treated group as compared with control group, Table S2: DEG annotation, Table S3: Transcription factors related to plant hormone signal transduction, stress response, and secondary metabolite synthesis.

Author Contributions: J.Z. designed the study, performed the experiments and data analysis, wrote and revised the manuscript; H.Y. provided some research literature of Wuzhimaotao, and helped in the preparation of plant materials. All authors have read and agreed to the published version of the manuscript.

Funding: This work was funded by the National Natural Science Foundation of China (31971568), Key Special Project for Introduced Talents Team of Southern Marine Science and Engineering Guangdong Laboratory (Guangzhou) (GML2019ZD0408), the Natural Science Foundation of Guangdong Province (2014A030313769; 2020A1515010540), and Key tasks of Guangdong rural science and Technology Commissioner in 2020 (KTP20200048).

Institutional Review Board Statement: Not applicable.

Informed Consent Statement: Not applicable.

Data Availability Statement: Data is contained within the article or supplementary material.

Conflicts of Interest: The authors declare no conflict of interest.

References

1. Song, L.R.; Hong, X.; Ding, X.L.; Zang, Z.Y. *Modern Chinese Pharmacy Dictionary*; People's Health Publishing House: Beijing, China, 2001.
2. Pharmacopoeia Committee of the Ministry of Health of the People's Republic of China. *Pharmacopoeia of the People's Republic of China*; People's Health Publishing House: Beijing, China, 1997.
3. Editorial Committee of Chinese Materia Medica; State Administration of Traditional Chinese Medicine. *Chinese Materia Medica*; Shanghai Science and Technology Press: Shanghai, China, 1999; Volume II.
4. Lin, H.; Mei, Q.X.; Zeng, C.Y. A survey of studies on chemical constituents and pharmacological activities of Wuzhimaotao (*Ficus hirta Vahl*). *Pharm. Today* **2012**, *22*, 484–486.
5. Huang, L.Q.; Guo, L.P. Secondary metabolites accumulating and geoherbs formation under environmental stress. *China J. Tradit. Chin. Mater. Med.* **2007**, *32*, 277–280.
6. Kuć, J.; Rush, J.S. Phytoalexins. *Arch. Biochem. Biophys.* **1985**, *236*, 455–472. [CrossRef]
7. Zeng, J.; Zhang, M.; Sun, X. Molecular hydrogen is involved in phytohormone signaling and stress responses in plants. *PLoS ONE* **2013**, *8*, e71038. [CrossRef] [PubMed]
8. Cui, W.; Fang, P.; Zhu, K.; Mao, Y.; Gao, C.; Xie, Y.; Wang, J.; Shen, W. Hydrogen-rich water confers plant tolerance to mercury toxicity in alfalfa seedlings. *Ecotoxicol. Environ. Saf.* **2014**, *105*, 103–111. [CrossRef]
9. Cui, W.; Gao, C.; Fang, P.; Lin, G.; Shen, W. Alleviation of cadmium toxicity in Medicago sativa by hydrogen-rich water. *J. Hazard. Mater.* **2013**, *260*, 715–724. [CrossRef]
10. Dai, C.; Cui, W.; Pan, J.; Xie, Y.; Wang, J.; Shen, W. Proteomic analysis provides insights into the molecular bases of hydrogen gas-induced cadmium resistance in Medicago sativa. *J. Proteom.* **2017**, *152*, 109–120. [CrossRef]
11. Zeng, J.; Ye, Z.; Sun, X. Progress in the study of biological effects of hydrogen on higher plants and its promising application in agriculture. *Med. Gas Res.* **2014**, *4*, 15. [CrossRef]
12. Renwick, G.M.; Giumarro, C.; Siegel, S.M. Hydrogen metabolism in higher plants. *Plant Physiol.* **1964**, *39*, 303–306. [CrossRef]
13. Dole, M.; Wilson, F.R.; Fife, W.P. Hyperbaric hydrogen therapy: A possible treatment for cancer. *Science* **1975**, *190*, 152–154. [CrossRef]
14. Gharib, B.; Hanna, S.; Abdallahi, O.M.; Lepidi, H.; Gardette, B.; De Reggi, M. Anti-inflammatory properties of molecular hydrogen: Investigation on parasite-induced liver inflammation. *C. R. Acad. Sci. III* **2001**, *324*, 719–724. [CrossRef]
15. Ohsawa, I.; Ishikawa, M.; Takahashi, K.; Watanabe, M.; Nishimaki, K.; Yamagata, K.; Katsura, K.; Katayama, Y.; Asoh, S.; Ohta, S. Hydrogen acts as a therapeutic antioxidant by selectively reducing cytotoxic oxygen radicals. *Nat. Med.* **2007**, *13*, 638–694. [CrossRef] [PubMed]
16. Li, C.; Gong, T.; Bian, B.; Liao, W. Roles of hydrogen gas in plants: A review. *Funct. Plant Biol.* **2018**, *45*, 783–792. [CrossRef]
17. Russell, G.; Zulfiqar, F.; Hancock, J.T. Hydrogenases and the Role of Molecular Hydrogen in Plants. *Plants* **2020**, *9*, 1136. [CrossRef]
18. Liu, F.; Li, J.; Liu, Y. Molecular hydrogen can take part in phytohormone signal pathways in wild rice. *Biol. Plant.* **2016**, *60*, 311–319. [CrossRef]
19. Yang, L.; Wen, K.S.; Ruan, X.; Zhao, Y.X.; Wei, F.; Wang, Q. Response of Plant Secondary Metabolites to Environmental Factors. *Molecules* **2018**, *23*, 762. [CrossRef]
20. Ramakrishna, A.; Ravishankar, G.A. Influence of abiotic stress signals on secondary metabolites in plants. *Plant Signal. Behav.* **2011**, *6*, 1720–1731. [CrossRef]
21. Ho, T.T.; Murthy, H.N.; Park, S.Y. Methyl Jasmonate Induced Oxidative Stress and Accumulation of Secondary Metabolites in Plant Cell and Organ Cultures. *Int. J. Mol. Sci.* **2020**, *21*, 716. [CrossRef]
22. Chen, J. The Research on the Bioactive Constituents of the Roots of Hairy Fig (*Ficus hirta Vahl.*). Master's Thesis, Guangdong Pharmaceutical University, Guangzhou, China, 2017.
23. Ouyang, G.; Xue, Y. Physiological role and regulation of phenylpropanoid metabolism in plant. *Plant Physiol. Commun.* **1988**, *3*, 9–16.
24. Sharma, A.; Shahzad, B.; Rehman, A.; Bhardwaj, R.; Landi, M.; Zheng, B. Response of Phenylpropanoid Pathway and the Role of Polyphenols in Plants under Abiotic Stress. *Molecules* **2019**, *24*, 2452. [CrossRef]
25. Dixon, R.A.; Achnine, L.; Kota, P.; Liu, C.J.; Reddy, M.S.; Wang, L. The phenylpropanoid pathway and plant defence—A genomics perspective. *Mol. Plant Pathol.* **2002**, *3*, 371–390. [CrossRef]
26. Korkina, L. Phenylpropanoids as naturally occurring antioxidants: From plant defense to human health. *Cell Mol. Biol.* **2007**, *53*, 15–25. [PubMed]
27. Lei, Z.; Zhou, C.; Ji, X.; Wei, G.; Huang, Y.; Yu, W.; Luo, Y.; Qiu, Y. Transcriptome Analysis Reveals genes involved in flavonoid biosynthesis and accumulation in Dendrobium catenatum From Different Locations. *Sci. Rep.* **2018**, *8*, 6373. [CrossRef] [PubMed]
28. Jiang, T.; Guo, K.; Liu, L.; Tian, W.; Xie, X.; Wen, S.; Wen, C. Integrated transcriptomic and metabolomic data reveal the flavonoid biosynthesis metabolic pathway in *Perilla frutescens* (L.) leaves. *Sci. Rep.* **2020**, *10*, 16207. [CrossRef]

29. Meraj, T.A.; Fu, J.; Raza, M.A.; Zhu, C.; Shen, Q.; Xu, D.; Wang, Q. Transcriptional Factors Regulate Plant Stress Responses through Mediating Secondary Metabolism. *Genes* **2020**, *11*, 346. [CrossRef] [PubMed]

30. Ng, D.W.; Abeysinghe, J.K.; Kamali, M. Regulating the Regulators: The Control of Transcription Factors in Plant Defense Signaling. *Int. J. Mol. Sci.* **2018**, *19*, 3737. [CrossRef]

31. Seo, E.; Choi, D. Functional studies of transcription factors involved in plant defenses in the genomics era. *Brief Funct. Genom.* **2015**, *14*, 260–267. [CrossRef]

32. Li, F.; Hu, Y.; Shan, Y.; Liu, J.; Ding, X.; Duan, X.; Zeng, J.; Jiang, Y. Hydrogen-rich water maintains the color quality of fresh-cut Chinese water chestnut. *Postharvest Biol. Technol.* **2022**, *183*, 111743. [CrossRef]

33. Grabherr, M.G.; Haas, B.J.; Yassour, M.; Levin, J.Z.; Thompson, D.A.; Amit, I.; Adiconis, X.; Fan, L.; Raychowdhury, R.; Zeng, Q.; et al. Full-length transcriptome assembly from RNA-Seq data without a reference genome. *Nat. Biotechnol.* **2011**, *29*, 644–652. [CrossRef]

34. Young, M.D.; Wakefield, M.J.; Smyth, G.K.; Oshlack, A. Gene ontology analysis for RNA-seq: Accounting for selection bias. *Genome Biol.* **2010**, *11*, R14. [CrossRef]

35. Kanehisa, M.; Araki, M.; Goto, S.; Hattori, M.; Hirakawa, M.; Itoh, M.; Katayama, T.; Kawashima, S.; Okuda, S.; Tokimatsu, T.; et al. KEGG for linking genomes to life and the environment. *Nucleic Acids Res.* **2008**, *36*, D480–D484. [CrossRef] [PubMed]

36. Mao, X.; Cai, T.; Olyarchuk, J.G.; Wei, L. Automated genome annotation and pathway identification using the KEGG Orthology (KO) as a controlled vocabulary. *Bioinformatics* **2005**, *21*, 3787–3793. [CrossRef] [PubMed]

Review

Downstream Signalling from Molecular Hydrogen

John T. Hancock * and Grace Russell

Department of Applied Sciences, University of the West of England, Bristol BS16 1QY, UK;
Grace2.Russell@live.uwe.ac.uk
* Correspondence: john.hancock@uwe.ac.uk; Tel.: +44-(0)-1173-282-475

Abstract: Molecular hydrogen (H_2) is now considered part of the suite of small molecules that can control cellular activity. As such, H_2 has been suggested to be used in the therapy of diseases in humans and in plant science to enhance the growth and productivity of plants. Treatments of plants may involve the creation of hydrogen-rich water (HRW), which can then be applied to the foliage or roots systems of the plants. However, the molecular action of H_2 remains elusive. It has been suggested that the presence of H_2 may act as an antioxidant or on the antioxidant capacity of cells, perhaps through the scavenging of hydroxyl radicals. H_2 may act through influencing heme oxygenase activity or through the interaction with reactive nitrogen species. However, controversy exists around all the mechanisms suggested. Here, the downstream mechanisms in which H_2 may be involved are critically reviewed, with a particular emphasis on the H_2 mitigation of stress responses. Hopefully, this review will provide insight that may inform future research in this area.

Keywords: antioxidants; heme oxygenase; hydrogen gas; hydrogenase; hydroxyl radicals; molecular hydrogen; nitric oxide; reactive oxygen species

Citation: Hancock, J.T.; Russell, G. Downstream Signalling from Molecular Hydrogen. *Plants* **2021**, *10*, 367. https://doi.org/10.3390/plants10020367

Academic Editor: Szilvia Z. Tóth

Received: 25 January 2021
Accepted: 9 February 2021
Published: 14 February 2021

Publisher's Note: MDPI stays neutral with regard to jurisdictional claims in published maps and institutional affiliations.

1. Introduction

Molecular hydrogen (H_2) is now recognized to have biochemical effects in both animals [1,2] and plants [3,4]. Although it is a relatively inert gas, H_2 appears to have profound effects on cell activity, which can be harnessed to help plant growth, survival, and productivity [5–8].

Plants, particularly as they are sessile, have to endure and survive a wide range of stress challenges, both biotic and abiotic. These stresses include attack by pathogens [9] and insects [10], as well as heavy metals [11], extreme temperature [12], salt [13], and ultraviolet B light [14]. It has become apparent over many years of study that there are common molecular responses to such stresses, and these mechanisms often involve reactive oxygen species (ROS) [15] and reactive nitrogen species (RNS) [16]. These compounds include ROS such as superoxide anions ($O_2 \cdot^-$) and hydrogen peroxide (H_2O_2), the latter of which is a major focus of ROS signalling [17]. Importantly, ROS also include the hydroxyl radical ($\cdot OH$). The most prominent RNS is nitric oxide (NO), which is known to be involved in plant cell signalling processes [18]. However, other RNS include peroxynitrite and nitrosoglutathione, both of which can act as signalling molecules [19,20]. It is also apparent that crosstalk occurs between ROS and RNS [21] as well as with other reactive signalling molecules such as hydrogen sulphide (H_2S) [22,23].

H_2 fits into this suite of reactive signalling molecules and was shown to increase the fitness of plants [24]. Suitable examples of recent papers on H_2 effects on plants include mitigation of salinity effects in barley [25] and Arabidopsis [26], and increased tolerance to cadmium in alfalfa [27]. However, exactly how H_2 interacts and has an effect is unclear. The metabolism of H_2 in plants is not a novel idea [28] and some plants are known to be significant generators of H_2, such as Chlamydomonas [29,30], whilst higher plants have been shown to produce H_2 too. Plant H_2 generation has been known for a long time [28,31], with more recent examples being reported using rice seedlings [32] and tomato plants [33]. The role of hydrogenase enzymes and the generation of H_2 by plants was recently reviewed [7].

Molecular hydrogen, being a gas, is hard to use either in laboratory or environmental settings. It is extremely flammable [34], relatively insoluble [35,36], and will readily move to the gas phase. Despite this, treatment with H_2 is often facilitated by the production of hydrogen-rich water (HRW), which can then be applied to the soil or directly onto the foliage. If using hydroponics, the HRW can be added directly to the feed solution. Several examples of the use of HRW are included throughout this review (for example, [5,8,37]). The use of HRW is effective and easy and is commonly used to treat plants, but treatment with H_2 gas can also have cellular effects and is often used in animal studies, for example, with mice [38]. H_2 gas has been used to alter plant growth by the gaseous treatment of the soil [39]. The treatment of biological materials with H_2 was further discussed in previous papers [7,40].

Here, we provide a critical look at the correlation between the effect of H_2 and the possible modes of action, with stress responses in plants being a focus. Issues that are addressed here include both the direct and indirect actions of H_2 and what biological compounds H_2 interacts within a cell, leading to the observed responses. Once this is established, a clearer view of downstream signal transduction initiated by H_2 can be gained. It is hoped that this review will inform future research in this area of plant science.

2. Downstream Effects

For any molecule to be used in cell signalling, it needs to be perceived by cells and to initiate a response. For many molecules, this involves a receptor protein, which may be on the cell surface [41] or in an intracellular compartment, such as the cytoplasm [42] or nucleus [43]. Some signalling molecules are perceived by proteins not classed as receptors, such as the effect of NO on soluble guanylyl cyclase (sGC). Here, NO reacts with the iron in the heme group of the enzyme, thereby activating it [44], although the involvement of such mechanisms has been questioned in plants [45]. Alternatively, the reactive nature of ROS and RNS allows them to oxidize [46] and nitrosate [47] thiol groups on proteins, propagating the signalling needed. It is hard to envisage how H_2, being so small and relatively inert, can be perceived by cells. Some of the mechanisms reported and mooted are discussed below.

2.1. Effects on Reactive Oxygen Species and Antioxidant Capacity

Stress responses in plants often involve ROS metabolism. There is often an increase in ROS accumulation, which, in some cases, can initiate programmed cell death (PCD) in plants [48]. ROS accumulate in the presence of heavy metals [49], such as cadmium [50], mercury, and copper [51]. ROS also accumulate in the presence of salt, extreme temperature, and pathogens [52]. Increases in the intracellular ROS under such stress conditions are often accompanied by an increase in antioxidant levels in cells, for example, in the presence of salt [53], heavy metals [54], and extreme temperature [55]. Therefore, the modulation of ROS metabolism is crucial for stress responses: increases in ROS lead to changes in cellular function, whilst antioxidants modulate and dampen that response.

H_2 has been shown to be able to help plant cells mitigate stress challenge. H_2 can help reduce salt stress [56,57], and reduce stress due to aluminium [58,59], cadmium [60], and mercury [61]. H_2 also can help mitigate against drought stress [62,63] and paraquat induced oxidative stress [64].

Xie et al. [57] suggested that H_2 modulates plant cells' antioxidant capacity through acting through zinc-finger transcription factor ZAT10/12. This would dampen the ROS accumulation and associated lipid peroxidation. They also suggested that H_2 would act on the antiporters and proton pumps responsible for exclusion of Na^+, particularly the protein salt overly sensitive1 (SOS1). Finally, it was suggested that both SOS1 and cytosolic ascorbate peroxidase1 (cAPX1) are molecular targets of H_2-mediated signalling. Additionally, Xu et al. [59] also suggested that H_2 may alter gene expression. In a study of aluminium stress, they found that H_2 altered the ratio of gibberellin acid (GA) and abscisic acid (ABA), with the expression of genes for GA biosynthesis (*GA20ox1* and *GA20ox2*) and for ABA breakdown (*ABA8ox1* and *ABA8ox2*) being induced by H_2. H_2 also altered miRNA expression with downstream effects that increased superoxide dismutase (SOD) expression, increasing antioxidant levels in the cells. However, even though these findings all support the notion that H_2 is protecting the cells, no direct interaction with H_2 has been established.

As can be seen from the discussion above, both stress responses and the effects of H_2 can be linked to ROS metabolism and antioxidant levels in cells. Therefore, it is particularly pertinent that H_2 has been posited to be an antioxidant [65]. Although this study discusses the effects in H_2 in a clinical setting, the redox chemistry would be the same in plants cells. In an animal setting, a study showed that H_2 is an antioxidant against the hydroxyl radical ($\cdot$OH) but has no effects against other ROS [66]. This is most significant, as it is usually hydrogen peroxide (H_2O_2) that is deemed to be the primary inter- and intracellular signal [17,67]. Of importance, the specificity of H_2 to scavenge $\cdot$OH has been disputed, as an in vitro study showed that H_2 can scavenge H_2O_2. However, H_2 could not scavenge superoxide anions [57]. In an experiment looking at the radiolysis of water, a negligible effect on the formation or consumption of H_2O_2 was seen when molecular hydrogen was added [68].

If, as suggested [66], the effects of H_2 are mediated partly by $\cdot$OH scavenging, a series of questions could be asked: How influential are the levels of hydroxyl radicals in cells, and could H_2 be acting through their modulation? Would this account for the effects seen?

Hydroxyl radicals are known to have effects in plant cells. Richards et al. [69] described the hydroxyl radical as being a "potent regulator in plant cell biology". They discussed the role of this molecule in numerous physiological mechanisms in plants, including germination, control of stomatal apertures, reproduction, and adaptation to stress challenge. $\cdot$OH has also been shown to be important for ion currents in roots [70,71]. In animal cells, $\cdot$OH was shown to be upstream of mitogen-activated protein kinases (MAPKs) and transcription factors (ERK2 and NF-κB) [72], and analogous mechanisms could exist in plants. Therefore, evidence exists of $\cdot$OH acting in a positive cell signalling role, which could potentially be the target of H_2.

In biological systems, ROS are often the product of the sequential reduction of molecular oxygen, resulting ultimately in the 4-electron reduction to water (Equation (1)).

$$O_2 \xrightarrow{e^-} O_2{}^- \xrightarrow{e^-} H_2O_2 \xrightarrow{e^-} 2(;OH) \xrightarrow{e^-} 2H_2O \tag{1}$$

The superoxide anion ($O_2\cdot^-$) can be produced enzymatically, for example from the action of NADPH oxidases [73]. H_2O_2 can be produced by the subsequent dismutation of $O_2\cdot^-$ by the enzyme family of superoxide dismutases (SOD) [74].

$\cdot$OH can be then be subsequently produced, especially in the presence of metal ions [75,76]. This generation can be either from the Fenton reaction from H_2O_2 (Equation (2)):

$$H_2O_2 + Fe^{2+} \rightarrow \cdot OH + HO^- + Fe^{3+} \tag{2}$$

Or in the presence of transition metals through the Haber–Weiss reaction, using superoxide anions and H_2O_2 (Equation (3)):

$$H_2O_2 + O_2{}^- \rightarrow \cdot OH + OH^- + O_2 \tag{3}$$

If the production of ROS is initiated, for example, during a stress response as discussed above, the generation of $\cdot$OH is likely to proceed. Hydroxyl radicals can be detected in plant cells [77,78], and have been found to have multiple effects.

The application of H_2 has mitigating influences during stress, and therefore if the effects of H_2 are mediated by the removal of $\cdot$OH, then it might be expected that $\cdot$OH radicals would need to be produced during these stress responses, assuming H_2 is working in these cases as a $\cdot$OH scavenger. It is in fact the case that $\cdot$OH can be found in these circumstances. For example, hydroxyl radicals increase during metal ion challenge [79], a cellular challenge in which H_2 has been shown to have a beneficial effect [58–61]. In a similar manner ·OH is produced during paraquat treatment of plants [80], another situation mitigated by H_2 [64]. During chilling stress and drought stress, increases in free iron and H_2O_2 have been recorded, and this implicates hydroxyl radical generation in downstream cellular responses [81]. Once again, H_2 has beneficial effects under drought conditions [62,63], as well as chilling stress [82]. ·OH and H_2 also have similar actions in heat stress [83,84]. Therefore, it can be seen that there are many stress conditions which elicit accumulation of ·OH and are also relieved by the presence of H_2, suggesting that the ·OH scavenging activity of H_2 is potentially responsible for the changes in cellular activity seen. This of course does not consider any spatial-temporal differences in ·OH accumulation during different stresses, or plant species variations, but the correlation of ·OH action and H_2 effects may be pointing to a possible mechanism.

Certainly, to support the notion that ·OH removal by H_2 could be biologically significant, a look at other ·OH scavengers may be useful. Such scavenging has been suggested to be useful for animal health [85], whilst in plants, mannitol has been suggested to be protective through this mechanism [81]. Sugars such as sucralose has been studied for its ·OH scavenging effects in Arabidopsis [86], whilst β-carboline alkaloids [87] and more novel compounds have been used in animal systems [88]. Such studies show that there is merit in modulating ·OH in cells, and therefore support the notion that such action by H_2 may be significant.

On the other hand, and importantly, it has been suggested that the reaction of H_2 with ·OH is too slow to be of physiological relevance [89], although the authors were discussing clinical settings. In this paper the rate constant for the reaction of H_2 with ·OH producing H_2O and H· is only 4.2×10^7 M^{-1} s^{-1} (from [90,91]). The rate constant for other radical reactions was quoted as 10^9 M^{-1} s^{-1}. It was suggested [89] that the ·OH would react with other biomolecules before reacting with the H_2, rendering the presence of H_2 as being irrelevant. Others have doubted whether H_2 has its effects through scavenging ·OH, although this is from a human health perspective [92]. Assuming this is correct, the correlation of ·OH production and H_2 effects during stress responses would also be irrelevant, begging the question, if ·OH scavenging is not the mechanism, what is?

It is possible that H_2 has indirect effects on antioxidant levels. There are several reports of antioxidant levels in plant cells altering on H_2 treatment. For example, this was reported in a study using black barley (*Hordeum distichum* L.) [93]. Antioxidant enzymes such as catalase and SOD were increased in maize [94] with similar effects in Chinese cabbage [95]. HRW was also found to maintain the intracellular redox status of plant cells through alterations the levels of reduced and oxidized glutathione (GSH and GSSG) [60]. However, the direct targets of H_2 have not been identified in such studies. Therefore, it may be that H_2 is having effects on the cells' antioxidant capacity, which can be measured, but it may not be a direct effect on the ROS themselves.

2.2. Impact on Reactive Nitrogen Species Metabolism

RNS, such as the nitric oxide radical (NO), have been known to have important effects in plant cells for over forty years [96], although there is still some controversy of their endogenous production and action [45]. NO, like ROS are well known to be involved in plant stress responses [97], many of which are ameliorated by H_2 treatment, as discussed above. Therefore, the relationship between H_2 presence and altered RNS metabolism is worth exploring.

H_2 has been shown to have effects in nitrogen fixation [98], although this is only one facet of this complex process. Nitrogen fixation relies on many factors including nutrient availability, the soil-plant interactions, and community facilitation as exemplified by the work carried out with the alpine shrub *Salix herbacea* [99–101]. H_2 has also been shown to alter NO synthesis during auxin-mediated root growth [33]. Li et al. [102] reported that NO was involved in H_2-induced root growth, whilst Zhu et al. [103] also link H_2 and NO, reporting that H_2 promoted NO accumulation through increases in the activities of possible synthesizing enzymes: NO synthase-like enzymes and nitrate reductase. Additionally, HRW increased NO accumulation in a study on stomatal closure [104]. On the other hand, HRW decreased NO accumulation in alfalfa [59].

It is likely that during a stress response NO and ROS are produced temporally and spatially together, and they can interact to produce downstream products. Superoxide anions and NO together can lead to the generation of the $\cdot OH$ radical [105], and as discussed above this have been mooted as a potential mechanism of H_2 action. However, superoxide anions and NO can react to produce peroxynitrite ($ONOO^-$) [105], which can act as a signalling molecule in its own right [106,107], possibility through alterations of amino acids [108], with tyrosine nitration being a major covalent change seen [106] which could have important downstream effects [109].

It has been reported that H_2 reacts with $ONOO^-$, but not NO [66,110]. Therefore, it would be unlikely that H_2 has direct effects in the NO signalling, *per se*. However, it was reported that H_2 reacts with peroxynitrite, which would potentially alter NO-induced signalling pathways. Despite several papers discussing the scavenging of $ONOO^-$ by H_2 [58,60], it has been completely ruled out by others [89]. In this paper, as well as saying that the $\cdot OH$ reaction is too slow, they report that H_2: (1) does not alter the rate of conversion of ONOOH to NO_3^- and H^+; (2) does not alter the rates of $ONOO^-$-mediated tyrosine nitration; (3) does not alter the oxidative stress responses mediated by either $ONOO^-$ or $\cdot OH$. Therefore, even if effects on NO metabolism are seen, such as alterations in activities of synthesising enzymes, there appears to be no direct scavenging of RNS, or $\cdot OH$, by H_2 which could account for the observed cellular effects.

2.3. Stress, Heme Oxygenase and H_2

An enzyme mechanism that has been found to be important for H_2 effects in cells involves the heme oxygenase enzyme (HO-1). For example, this was shown to be involved in root development in cucumber on treatment with HRW [37]. Hydrogen-mediated tolerance to paraquat was also shown to involve heme oxygenase [64]. Similar data can be found in studies of animal systems, for example, in mice [111].

HO-1 has been shown to be involved in a range of abiotic stress responses in plants, including salt, heavy metals, UV light, and drought. Responses to stresses such as drought are complex, involving the result of many genes being expressed and the effects of gene polymorphisms, as seen with *Phaseolus vulgaris* L. [112–115], with wild types showing tolerance differences [116,117]. Resistance and tolerance to extreme temperatures are also important and involve complicated cellular responses [118–121]. Such responses are often associated with the accumulation of cellular ROS and RNS [120]. The catalytic action of HO-1 is the breakdown of heme. This is an oxygen-dependent reaction that uses NADPH as a cofactor and generates biliverdin, carbon monoxide (CO), and iron [121,122]. Interestingly, CO has been shown to be involved in signalling events in cells, and could mediate downstream effects of H_2, whilst iron facilitates $\cdot OH$ production, as discussed above.

However, no direct interaction between H_2 and HO-1 seems to have been reported. Further, no reaction has been reported between H_2 and CO in biological systems. Therefore, the connection between H_2 treatment and alterations of HO-1 activity needs to be a focus for future research.

2.4. Paramagnetic Properties and Possible Cellular Effects

The above discussion throws doubt onto many biochemical and reactive aspects of H_2 effects in cells. However, the physical properties of H_2 may also be important. Hydrogen can exist with two nuclear spin states (ortho- and parahydrogen) [123,124]. It is the inter-conversion between these states that may be relevant here [125]. One of the interactions discussed was with NO, which could potentially alter NO signalling. There is also the possibility of interactions with transition metals [126]. This could have a potentially significant effect on cell signalling pathways, as many enzymes involved in signal transduction have metal prosthetic groups, including guanylyl cyclase (at least in animals), SOD, and many respiratory and photosynthetic components. Many of the aforementioned enzymes may be involved in ROS and RNS metabolism, which are important in plant responses to many stresses, with such conditions being mitigated by H_2, as discussed above. It is conceivable that H_2 may interact with the heme during the catalytic cycle of HO-1, accounting for the effects mediated by this enzyme.

This physical aspect of H_2 action was mooted previously [127], although experimental evidence is lacking and future research may prove this avenue wrong. However, the idea of quantum biology is not confined to H_2 effects, and the topic was recently reviewed [128]. It was suggested that biological processes may occur due to quantum mechanical effects. A more recent review on this topic was also published [129].

3. Discussion

H_2 is known to be involved in the control of cellular functions in plant cells. For example, it was reported to be involved in both phytohormone signalling and stress responses [32]. On a pragmatic note, treatment with H_2 in the form of HRW was suggested to be useful for delaying postharvest spoilage of fruit [5]. Therefore, it is known, like animal cells [1,130], that H_2 has effects, and such actions may be harnessed for future manipulation of plant growth and crop enhancement [131].

Several mechanisms of H_2 action have been suggested, as summarized in Figure 1.

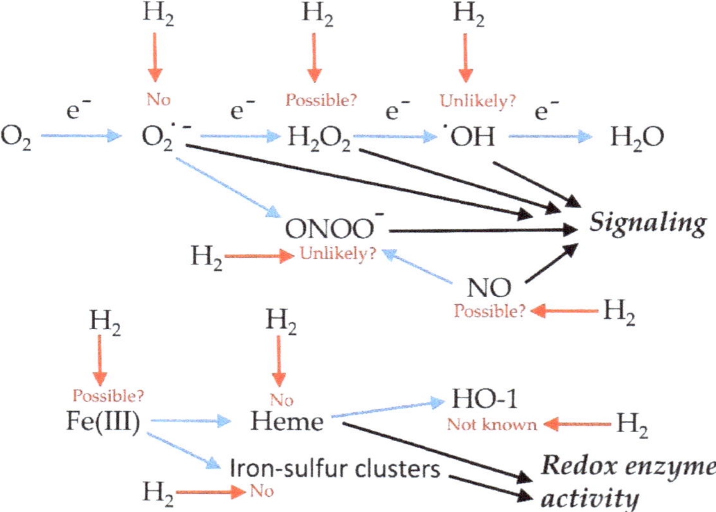

Figure 1. Possible mechanism of action of H_2 in cells. The likelihood of there being effects on particular molecules is indicated (red arrows and text).

One of the significant actions of H_2 in biological systems was suggested to be its $\cdot$OH scavenging activity [66], as reported in animal systems [132]. A range of studies have shown that $\cdot$OH increases in cells under stressful conditions [79–81], whilst H_2 has been shown to have effects on such stress responses [58–61]. It may be argued that removal of $\cdot$OH by H_2, if it is involved in important $\cdot$OH signalling pathways, should be detrimental to cell function, although many studies have looked at scavenging $\cdot$OH as a beneficial approach to cell and organism health, both in plants and animals [81,85–87]. Hydroxyl radicals are extremely reactive, and react with kinetics that are diffusion-limited, with rate constants for a range of biomolecules being determined, including ATP and ADP [133]. $\cdot$OH radicals are known to react with proteins [134], which can lead to amino acid oxidation, crosslinking, and degradation of the polypeptide [135]. Lipids [136], carbohydrates [137], and DNA [138] are also ·OH targets. Therefore, the scavenging activity of H_2 may prevent the harmful effects of ·OH, which may account for some of the observed effects. However, the biggest issue is the rate constant of the reaction between H_2 and ·OH, which is deemed to be too slow for physiological relevance [89], suggesting that the other biomolecules may react first anyway, and therefore H_2 would not influence the levels of oxidative stress. The same authors also ruled out reactions with peroxynitrite, as discussed above. Therefore, with H_2 not able to scavenge other ROS [66] and the effects of H_2 on both ·OH and $ONOO^-$ being ruled out [89], it appears that the scavenging role of H_2 may have limited effects in cells, at best.

Heme oxygenase is one enzyme that has been reported as mediating H_2 effects [37,64]. Although being reported in several studies, as discussed above, there is little evidence of a direct interaction which could account for the data seen. However, not all the data are negative and seemingly point to dead ends. It was reported that H_2 scavenged H_2O_2 [57], which, if confirmed and can be shown to have effects *in vivo*, would be very significant, as H_2O_2 is one of the major ROS signalling molecules [17,67]. However, in radiolysis experiments with H_2O_2, the addition of H_2 only had a negligible effect [68], suggesting that more research in this area would be beneficial. Another positive effect that is worth exploring is the interaction of H_2 with metals. It was suggested that the beneficial effects of H_2 may be mediated by the reduction of $Fe(III)$, oxidized as a result of oxidative stress. However, neither iron-sulphur clusters nor heme groups were reduced by the presence of H_2 [89]. Even so, the effect of H_2 on $Fe(III)$ is an enticing suggestion, as transition metals are widely used in biological systems, making this is another area that merits further investigation.

Finally, the paramagnetic properties of hydrogen may be relevant to its biological action, as previously mooted [127]. This may include interactions with NO or transition metals, but experimental data would be needed to support this notion. There are other papers with H_2 in catalysis, but it is difficult to determine their relevance to biochemical reactions, as they are often conducted under non-physiological conditions, such as high pressure [139].

In conclusion, although the involvement of molecular hydrogen in plant function has been known for a long time [28], there is still considerable uncertainty surrounding the exact actions of H_2 in cells. Its role as a direct antioxidant is doubted, although many cellular effects have been observed, including alterations in antioxidants, changes in enzyme activity, and modulation in gene expression. What is clear is that H_2 may be useful for the mitigation of plant stress, so it has been proposed to have an exciting future [4,131].

Author Contributions: J.T.H. wrote the draft of the manuscript and G.R. contributed to and edited the final version. All authors have read and agreed to the published version of the manuscript.

Funding: This research received no external funding.

Data Availability Statement: There were no primary data generated in this study.

Conflicts of Interest: The authors declare no conflict of interest.

References

1. Ge, L.; Yang, M.; Yang, N.-N.; Yin, X.-X.; Song, W.-G. Molecular hydrogen: A preventive and therapeutic medical gas for various diseases. *Oncotarget* **2017**, *8*, 102653–102673. [CrossRef]
2. Huang, L. Molecular hydrogen: A therapeutic antioxidant and beyond. *Med. Gas Res.* **2016**, *6*, 219–222. [CrossRef]
3. Li, C.; Gong, T.; Bian, B.; Liao, W. Roles of hydrogen gas in plants: A review. *Funct. Plant Biol.* **2018**, *45*, 783–792. [CrossRef]
4. Zeng, J.; Ye, Z.; Sun, X. Progress in the study of biological effects of hydrogen on higher plants and its promising application in agriculture. *Med. Gas Res.* **2014**, *4*, 1–4. [CrossRef]
5. Hu, H.; Li, P.; Wang, Y.; Gu, R. Hydrogen-rich water delays postharvest ripening and senescence of kiwifruit. *Food Chem.* **2014**, *156*, 100–109. [CrossRef]
6. Wilson, H.R.; Veal, D.; Whiteman, M.; Hancock, J.T. Hydrogen gas and its role in cell signalling. *CAB Rev.* **2017**, *12*, 1–3. [CrossRef]
7. Russell, G.; Zulfiqar, F.; Hancock, J.T. Hydrogenases and the Role of Molecular Hydrogen in Plants. *Plants* **2020**, *9*, 1136. [CrossRef] [PubMed]

8. Wu, Q.; Su, N.; Huang, X.; Ling, X.; Yu, M.; Cui, J.; Shabala, S. Hydrogen-rich water promotes elongation of hypocotyls and roots in plants through mediating the level of endogenous gibberellin and auxin. *Funct. Plant Biol.* **2020**, *47*, 771. [CrossRef] [PubMed]
9. Dodds, P.N.; Rathjen, J.P. Plant immunity: Towards an integrated view of plant–pathogen interactions. *Nat. Rev. Genet.* **2010**, *11*, 539–548. [CrossRef]
10. Cory, J.S.; Myers, J.H. Adaptation in an insect host-plant pathogen interaction. *Ecol. Lett.* **2004**, *7*, 632–639. [CrossRef]
11. Keyster, M.; Niekerk, L.-A.; Basson, G.; Carelse, M.; Bakare, O.; Ludidi, N.; Klein, A.; Mekuto, L.; Gokul, A. Decoding Heavy Metal Stress Signalling in Plants: Towards Improved Food Security and Safety. *Plants* **2020**, *9*, 1781. [CrossRef]
12. Ding, Y.; Shi, Y.; Yang, S. Molecular Regulation of Plant Responses to Environmental Temperatures. *Mol. Plant* **2020**, *13*, 544–564. [CrossRef] [PubMed]
13. Yang, Y.; Guo, Y. Unraveling salt stress signaling in plants. *J. Integr. Plant Biol.* **2018**, *60*, 796–804. [CrossRef]
14. Vanhaelewyn, L.; Prinsen, E.; Van Der Straeten, D.; Vandenbussche, F. Hormone-controlled UV-B responses in plants. *J. Exp. Bot.* **2016**, *67*, 4469–4482. [CrossRef] [PubMed]
15. Choudhury, F.K.; Rivero, R.M.; Blumwald, E.; Mittler, R. Reactive oxygen species, abiotic stress and stress combination. *Plant J.* **2016**, *90*, 856–867. [CrossRef]
16. Sharma, A.; Soares, C.; Sousa, B.; Martins, M.; Kumar, V.; Shahzad, B.; Sidhu, G.P.; Bali, A.S.; Asgher, M.; Bhardwaj, R.; et al. Nitric oxide-mediated regulation of oxidative stress in plants under metal stress: A review on molecular and biochemical aspects. *Physiol. Plant.* **2019**, *168*, 318–344. [CrossRef] [PubMed]
17. Černý, M.; Habánová, H.; Berka, M.; Luklová, M.; Brzobohatý, B. Hydrogen Peroxide: Its Role in Plant Biology and Crosstalk with Signalling Networks. *Int. J. Mol. Sci.* **2018**, *19*, 2812. [CrossRef] [PubMed]
18. Umbreen, S.; Lubega, J.; Cui, B.; Pan, Q.; Jiang, J.; Loake, G.J. Specificity in nitric oxide signalling. *J. Exp. Bot.* **2018**, *69*, 3439–3448. [CrossRef]
19. Speckmann, B.; Steinbrenner, H.; Grune, T.; Klotz, L.-O. Peroxynitrite: From interception to signaling. *Arch. Biochem. Biophys.* **2016**, *595*, 153–160. [CrossRef]
20. Ventimiglia, L.; Mutus, B. The Physiological Implications of S-Nitrosoglutathione Reductase (GSNOR) Activity Mediating NO Signalling in Plant Root Structures. *Antioxidants* **2020**, *9*, 1206. [CrossRef]
21. Lindermayr, C. Crosstalk between reactive oxygen species and nitric oxide in plants: Key role of S-nitrosoglutathione reductase. *Free. Radic. Biol. Med.* **2018**, *122*, 110–115. [CrossRef]
22. Hancock, J.; Whiteman, M. Hydrogen sulfide and cell signaling: Team player or referee? *Plant Physiol. Biochem.* **2014**, *78*, 37–42. [CrossRef]
23. Shivaraj, S.M.; Vats, S.; Bhat, J.A.; Dhakte, P.; Goyal, V.; Khatri, P.; Kumawat, S.; Singh, A.; Prasad, M.; Sonah, H.; et al. Nitric oxide and hydrogen sulfide crosstalk during heavy metal stress in plants. *Physiol. Plant.* **2019**, *168*, 437–455. [CrossRef] [PubMed]
24. Liu, F.; Jiang, W.; Han, W.; Li, J.; Liu, Y. Effects of Hydrogen-Rich Water on Fitness Parameters of Rice Plants. *Agron. J.* **2017**, *109*, 2033–2039. [CrossRef]
25. Wu, Q.; Su, N.; Shabala, L.; Huang, L.; Yu, M.; Shabala, S. Understanding the mechanistic basis of ameliorating effects of hydrogen rich water on salinity tolerance in barley. *Environ. Exp. Bot.* **2020**, *177*, 104136. [CrossRef]
26. Su, J.; Yang, X.; Shao, Y.; Chen, Z.; Shen, W. Molecular hydrogen–induced salinity tolerance requires melatonin signalling in Arabidopsis thaliana. *Plant Cell Environ.* **2021**, *44*, 476–490. [CrossRef]
27. Cui, W.; Yao, P.; Pan, J.; Dai, C.; Cao, H.; Chen, Z.; Zhang, S.; Xu, S.; Shen, W. Transcriptome analysis reveals insight into molecular hydrogen-induced cadmium tolerance in alfalfa: The prominent role of sulfur and (homo)glutathione metabolism. *BMC Plant Biol.* **2020**, *20*, 19–58. [CrossRef] [PubMed]
28. Renwick, G.M.; Giumarro, C.; Siegel, S.M. Hydrogen Metabolism in Higher Plants. *Plant Physiol.* **1964**, *39*, 303–306. [CrossRef]
29. Vargas, S.R.; Dos Santos, P.V.; Giraldi, L.A.; Zaiat, M.; Calijuri, M.D.C. Anaerobic phototrophic processes of hydrogen production by different strains of microalgae *Chlamydomonas* sp. *FEMS Microbiol. Lett.* **2018**, *365*, fny073. [CrossRef]
30. Hemschemeier, A.; Fouchard, S.; Cournac, L.; Peltier, G.; Happe, T. Hydrogen production by Chlamydomonas reinhardtii: An elaborate interplay of electron sources and sinks. *Planta* **2007**, *227*, 397–407. [CrossRef]
31. Sanadze, G.A. Absorption of molecular hydrogen by green leaves in light. *Fiziol. Rast.* **1961**, *8*, 555–559.
32. Zeng, J.; Zhang, M.; Sun, X. Molecular Hydrogen Is Involved in Phytohormone Signaling and Stress Responses in Plants. *PLoS ONE* **2013**, *8*, e71038. [CrossRef]
33. Cao, Z.; Duan, X.; Yao, P.; Cui, W.; Cheng, D.; Zhang, J.; Jin, Q.; Chen, J.; Dai, T.; Shen, W. Hydrogen Gas Is Involved in Auxin-Induced Lateral Root Formation by Modulating Nitric Oxide Synthesis. *Int. J. Mol. Sci.* **2017**, *18*, 2084. [CrossRef]
34. Jin, T.; Liu, Y.; Wei, J.; Wu, M.; Lei, G.; Chen, H.; Lan, Y. Modeling and analysis of the flammable vapor cloud formed by liquid hydrogen spills. *Int. J. Hydrogen Energy* **2017**, *42*, 26762–26770. [CrossRef]
35. Molecular Hydrogen Institute. Concentration and Solubility of H2. Available online: http://www.molecularhydrogeninstitute.com/concentration-and-solubility-of-h2 (accessed on 13 January 2021).
36. Wilhelm, E.; Battino, R.; Wilcock, R.J. Low-pressure solubility of gases in liquid water. *Chem. Rev.* **1977**, *77*, 219–262. [CrossRef]
37. Lin, Y.; Zhang, W.; Qi, F.; Cui, W.; Xie, Y.; Shen, W. Hydrogen-rich water regulates cucumber adventitious root development in a heme oxygenase-1/carbon monoxide-dependent manner. *J. Plant Physiol.* **2014**, *171*, 1–8. [CrossRef]
38. Fang, S.; Li, X.; Wei, X.; Zhang, Y.; Ma, Z.; Wei, Y.; Wang, W. Beneficial effects of hydrogen gas inhalation on a murine model of allergic rhinitis. *Exp. Ther. Med.* **2018**, *16*, 5178–5184. [CrossRef] [PubMed]

39. Dong, Z.; Wu, L.; Kettlewell, B.; Caldwell, C.D.; Layzell, D.B. Hydrogen fertilization of soils–is this a benefit of legumes in rotation? *Plant Cell Environ.* **2003**, *26*, 1875–1879. [CrossRef]
40. Hancock, J.T. Methods for the addition of redox compounds. In *Redox-Mediated Signal Transduction*; Hancock, J.T., Conway, M.E., Eds.; Humana: New York, NY, USA, 2019; pp. 13–25.
41. Deller, M.C. Cell surface receptors. *Curr. Opin. Struct. Biol.* **2000**, *10*, 213–219. [CrossRef]
42. Gasc, J.M.; Baulieu, E.E. Steroid hormone receptors: Intracellular distribution. *Biol. Cell* **1986**, *56*, 1–6. [CrossRef]
43. Mazaira, G.I.; Zgajnar, N.R.; Lotufo, C.M.; Daneri-Becerra, C.; Sivils, J.C.; Soto, O.B.; Cox, M.B.; Galigniana, M.D. Nuclear Receptors: A Historical Perspective. *Adv. Struct. Saf. Stud.* **2019**, *1966*, 1–5. [CrossRef]
44. Sandner, P.; Zimmer, D.P.; Milne, G.T.; Follmann, M.; Hobbs, A.; Stasch, J.-P. Soluble Guanylate Cyclase Stimulators and Activators. In *Handbook of Experimental Pharmacology*; Springer: Berlin/Heidelberg, Germany, 2018; pp. 1–39. [CrossRef]
45. Astier, J.; Mounier, A.; Santolini, J.; Jeandroz, S.; Wendehenne, D. The evolution of nitric oxide signalling diverges between animal and green lineages. *J. Exp. Bot.* **2019**, *70*, 4355–4364. [CrossRef] [PubMed]
46. Smirnoff, N.; Arnaud, D. Hydrogen peroxide metabolism and functions in plants. *New Phytol.* **2018**, *221*, 1197–1214. [CrossRef]
47. Feng, J.; Chen, L.; Zuo, J. Protein S -Nitrosylation in plants: Current progresses and challenges. *J. Integr. Plant Biol.* **2019**, *61*, 1206–1223. [CrossRef]
48. Petrov, V.; Hille, J.; Mueller-Roeber, B.; Gechev, T.S. ROS-mediated abiotic stress-induced programmed cell death in plants. *Front. Plant Sci.* **2015**, *6*, 69. [CrossRef] [PubMed]
49. Fryzova, R.; Pohanka, M.; Martinkova, P.; Cihlarova, H.; Brtnicky, M.; Hladky, J.; Kynicky, J. Oxidative Stress and Heavy Metals in Plants. *Rev. Environ. Contam. Toxicol.* **2017**, *245*, 129–156. [CrossRef]
50. Anjum, S.A.; Tanveer, M.; Hussain, S.; Bao, M.; Wang, L.; Khan, I.; Ullah, E.; Tung, S.A.; Samad, R.A.; Shahzad, B. Cadmium toxicity in Maize (*Zea mays* L.): Consequences on antioxidative systems, reactive oxygen species and cadmium accumulation. *Environ. Sci. Pollut. Res.* **2015**, *22*, 17022–17030. [CrossRef]
51. Górska-Czekaj, M.; Borucki, W. A correlative study of hydrogen peroxide accumulation after mercury or copper treatment observed in root nodules of *Medicago truncatula* under light, confocal and electron microscopy. *Micron* **2013**, *52–53*, 24–32. [CrossRef]
52. Camejo, D.; Guzmán-Cedeño, Á.; Moreno, A. Reactive oxygen species, essential molecules, during plant–pathogen interactions. *Plant Physiol. Biochem.* **2016**, *103*, 10–23. [CrossRef]
53. Santos, A.D.A.; Da Silveira, J.A.G.; Bonifacio, A.; Rodrigues, A.C.; Figueiredo, M.D.V.B. Antioxidant response of cowpea co-inoculated with plant growth-promoting bacteria under salt stress. *Braz. J. Microbiol.* **2018**, *49*, 513–521. [CrossRef] [PubMed]
54. AbdelGawad, H.; Zinta, G.; Hamed, B.A.; Selim, S.; Beemster, G.; Hozzein, W.N.; Wadaan, M.A.; Asard, H.; Abuelsoud, W.; Badreldin, A.H. Maize roots and shoots show distinct profiles of oxidative stress and antioxidant defense under heavy metal toxicity. *Environ. Pollut.* **2020**, *258*, 113705. [CrossRef] [PubMed]
55. Airaki, M.; Leterrier, M.; Mateos, R.M.; Valderrama, R.; Chaki, M.; Barroso, J.B.; Del Río, L.A.; Palma, J.M.; Corpas, F.J. Metabolism of reactive oxygen species and reactive nitrogen species in pepper (*Capsicum annuum* L.) plants under low temperature stress. *Plant Cell Environ.* **2012**, *35*, 281–295. [CrossRef] [PubMed]
56. Xu, S.; Zhu, S.; Jiang, Y.; Wang, N.; Wang, R.; Shen, W.; Yang, J. Hydrogen-rich water alleviates salt stress in rice during seed germination. *Plant Soil* **2013**, *370*, 47–57. [CrossRef]
57. Xie, Y.; Mao, Y.; Lai, D.; Zhang, W.; Shen, W. H$_2$ enhances Arabidopsis salt tolerance by manipulating ZAT10/12-mediated antioxidant defence and controlling sodium exclusion. *PLoS ONE* **2012**, *7*, e49800. [CrossRef] [PubMed]
58. Chen, M.; Cui, W.; Zhu, K.; Xie, Y.; Zhang, C.; Shen, W. Hydrogen-rich water alleviates aluminium-induced inhibition of root elongation in alfalfa via decreasing nitric oxide production. *J. Hazard. Mater.* **2014**, *267*, 40–47. [CrossRef]
59. Xu, D.; Cao, H.; Fang, W.; Pan, J.; Chen, J.; Zhang, J.; Shen, W. Linking hydrogen-enhanced rice aluminum tolerance with the reestablishment of GA/ABA balance and miRNA-modulated gene expression: A case study on germination. *Ecotoxicol. Environ. Saf.* **2017**, *145*, 303–312. [CrossRef]
60. Cui, W.; Gao, C.; Fang, P.; Lin, G.; Shen, W. Alleviation of cadmium toxicity in Medicago sativa by hydrogen-rich water. *J. Hazard. Mater.* **2013**, *260*, 715–724. [CrossRef] [PubMed]
61. Cui, W.; Fang, P.; Zhu, K.; Mao, Y.; Gao, C.; Xie, Y.; Wang, J.; Shen, W. Hydrogen-rich water confers plant tolerance to mercury toxicity in alfalfa seedlings. *Ecotoxicol. Environ. Saf.* **2014**, *105*, 103–111. [CrossRef]
62. Chen, Y.; Wang, M.; Hu, L.; Liao, W.; Dawuda, M.M.; Li, C. Carbon monoxide is involved in hydrogen gas-induced adventitious root development in cucumber under simulated drought stress. *Front. Plant Sci.* **2017**, *8*, 128. [CrossRef]
63. Jin, Q.; Zhu, K.; Cui, W.; Li, L.; Shen, W. Hydrogen-Modulated Stomatal Sensitivity to Abscisic Acid and Drought Tolerance Via the Regulation of Apoplastic pH in Medicago sativa. *J. Plant Growth Regul.* **2016**, *35*, 565–573. [CrossRef]
64. Jin, Q.; Zhu, K.; Cui, W.; Xie, Y.; Han, B.; Shen, W. Hydrogen gas acts as a novel bioactive molecule in enhancing plant tolerance to paraquat-induced oxidative stress via the modulation of heme oxygenase-1 signalling system. *Plant Cell Environ.* **2012**, *36*, 956–969. [CrossRef]
65. Hong, Y.; Chen, S.; Zhang, J.-M. Hydrogen as a Selective Antioxidant: A Review of Clinical and Experimental Studies. *J. Int. Med. Res.* **2010**, *38*, 1893–1903. [CrossRef] [PubMed]

66. Ohsawa, I.; Ishikawa, M.; Takahashi, K.; Watanabe, M.; Nishimaki, K.; Yamagata, K.; Katsura, K.-I.; Katayama, Y.; Asoh, S.; Ohta, S. Hydrogen acts as a therapeutic antioxidant by selectively reducing cytotoxic oxygen radicals. *Nat. Med.* **2007**, *13*, 688–694. [CrossRef] [PubMed]

67. Veal, E.A.; Day, A.M.; Morgan, B.A. Hydrogen Peroxide Sensing and Signaling. *Mol. Cell* **2007**, *26*, 1–14. [CrossRef]

68. Pastina, B.; LaVerne, J.A. Effect of Molecular Hydrogen on Hydrogen Peroxide in Water Radiolysis. *J. Phys. Chem. A* **2001**, *105*, 9316–9322. [CrossRef]

69. Richards, S.L.; Wilkins, K.A.; Swarbreck, S.M.; Anderson, A.A.; Habib, N.; Smith, A.G.; McAinsh, M.; Davies, J.M. The hydroxyl radical in plants: From seed to seed. *J. Exp. Bot.* **2015**, *66*, 37–46. [CrossRef] [PubMed]

70. Pottosin, I.; Zepeda-Jazo, I.; Bose, J.; Shabala, S. An Anion Conductance, the Essential Component of the Hydroxyl-Radical-Induced Ion Current in Plant Roots. *Int. J. Mol. Sci.* **2018**, *19*, 897. [CrossRef]

71. Demidchik, V.; Cuin, T.A.; Svistunenko, D.; Smith, S.J.; Miller, A.J.; Shabala, S.; Sokolik, A.; Yurin, V. Arabidopsis root K+-efflux conductance activated by hydroxyl radicals: Single-channel properties, genetic basis and involvement in stress-induced cell death. *J. Cell Sci.* **2010**, *123*, 1468–1479. [CrossRef]

72. Lu, W.J.; Lin, K.H.; Hsu, M.J.; Chou, D.S.; Hsiao, G.; Sheu, J.R. Suppression of NF-κB signaling by andrographolide with a novel mechanism in human platelets: Regulatory roles of the p38 MAPK-hydroxyl radical-ERK2 cascade. *Biochem. Pharmacol.* **2012**, *84*, 914–924. [CrossRef]

73. Qu, Y.; Yan, M.; Zhang, Q. Functional regulation of plant NADPH oxidase and its role in signaling. *Plant Signal. Behav.* **2017**, *12*, e1356970. [CrossRef]

74. Gill, S.S.; Anjum, N.A.; Gill, R.; Yadav, S.; Hasanuzzaman, M.; Fujita, M.; Mishra, P.; Sabat, S.C.; Tuteja, N. Superoxide dismutase—Mentor of abiotic stress tolerance in crop plants. *Environ. Sci. Pollut. Res.* **2015**, *22*, 10375–10394. [CrossRef]

75. Fong, K.-L.; McCay, P.B.; Poyer, J.; Misra, H.P.; Keele, B.B. Evidence for superoxide-dependent reduction of Fe^{3+} and its role in enzyme-generated hydroxyl radical formation. *Chem. Interact.* **1976**, *15*, 77–89. [CrossRef]

76. Halliwell, B. Superoxide-dependent formation of hydroxyl radicals in the presence of iron chelates: Is it a mechanism for hydroxyl radical production in biochemical systems? *FEBS Lett.* **1978**, *92*, 321–326. [CrossRef]

77. Kumar, A.; Prasad, A.; Sedlářová, M.; Pospíšil, P. Data on detection of singlet oxygen, hydroxyl radical and organic radical in Arabidopsis thaliana. *Data Brief* **2018**, *21*, 2246–2252. [CrossRef] [PubMed]

78. Chen, W.; Ding, S.; Wu, J.; Shi, G.; Zhu, A. In situ detection of hydroxyl radicals in mitochondrial oxidative stress with a nanopipette electrode. *Chem. Commun.* **2020**, *56*, 13225–13228. [CrossRef] [PubMed]

79. Cuypers, A.; Hendrix, S.; Dos Reis, R.A.; De Smet, S.; Deckers, J.; Gielen, H.; Jozefczak, M.; Loix, C.; Vercampt, H.; Vangronsveld, J.; et al. Hydrogen Peroxide, Signaling in Disguise during Metal Phytotoxicity. *Front. Plant Sci.* **2016**, *7*, 470. [CrossRef]

80. Babbs, C.F.; Pham, J.A.; Coolbaugh, R.C. Lethal Hydroxyl Radical Production in Paraquat-Treated Plants. *Plant Physiol.* **1989**, *90*, 1267–1270. [CrossRef]

81. Shen, B.; Jensen, R.G.; Bohnert, H.J. Mannitol Protects against Oxidation by Hydroxyl Radicals. *Plant Physiol.* **1997**, *115*, 527–532. [CrossRef]

82. Liu, F.J.; Cai, B.B.; Sun, S.N.; Bi, H.G.; Ai, X.Z. Effect of hydrogen-rich water soaked cucumber seeds on cold tolerance and its physiological mechanism in cucumber seedlings. *Sci. Agric. Sin.* **2017**, *50*, 881–889.

83. Yadav, D.K.; Pospíšil, P. Role of chloride ion in hydroxyl radical production in photosystem II under heat stress: Electron paramagnetic resonance spin-trapping study. *J. Bioenerg. Biomembr.* **2012**, *44*, 365–372. [CrossRef]

84. Chen, Q.; Zhao, X.; Lei, D.; Hu, S.; Shen, Z.; Shen, W.; Xu, X. Hydrogen-rich water pretreatment alters photosynthetic gas exchange, chlorophyll fluorescence, and antioxidant activities in heat-stressed cucumber leaves. *Plant Growth Regul.* **2017**, *83*, 69–82. [CrossRef]

85. Lipinski, B. Hydroxyl Radical and Its Scavengers in Health and Disease. *Oxidative Med. Cell. Longev.* **2011**, *2011*, 1–9. [CrossRef] [PubMed]

86. Matros, A.; Peshev, D.; Peukert, M.; Mock, H.-P.; Ende, W.V.D. Sugars as hydroxyl radical scavengers: Proof-of-concept by studying the fate of sucralose in Arabidopsis. *Plant J.* **2015**, *82*, 822–839. [CrossRef] [PubMed]

87. Herraiz, T.; Galisteo, J. Hydroxyl radical reactions and the radical scavenging activity of β-carboline alkaloids. *Food Chem.* **2015**, *172*, 640–649. [CrossRef] [PubMed]

88. Sakai, T.; Imai, J.; Ito, T.; Takagaki, H.; Ui, M.; Hatta, S. The novel antioxidant TA293 reveals the role of cytoplasmic hydroxyl radicals in oxidative stress-induced senescence and inflammation. *Biochem. Biophys. Res. Commun.* **2017**, *482*, 1183–1189. [CrossRef]

89. Penders, J.; Kissner, R.; Koppenol, W.H. ONOOH does not react with H_2: Potential beneficial effects of H_2 as an antioxidant by selective reaction with hydroxyl radicals and peroxynitrite. *Free Radic. Biol. Med.* **2014**, *75*, 191–194. [CrossRef]

90. Buxton, G.V.; Greenstock, C.L.; Helman, P.; Ross, A.B. Critical-review of rate constants for reactions of hydrated electrons, hydrogen-atoms and hydroxyl radicals (OH/O^-) in aqueous-solution. *J. Phys. Chem. Ref. Data* **1988**, *17*, 513–886. [CrossRef]

91. Matheson, M.S.; Rabani, J. Pulse radiolysis of aqueous hydrogen solutions. I. Rate constants for reaction of eaq− with itself and other transients. II. The interconvertibility of eaq− and H. *J. Phys. Chem.* **1965**, *69*, 1324–1335. [CrossRef]

92. Wood, K.C.; Gladwin, M.T. The hydrogen highway to reperfusion therapy. *Nat. Med.* **2007**, *13*, 673–674. [CrossRef] [PubMed]

93. Guan, Q.; Ding, X.-W.; Jiang, R.; Ouyang, P.-L.; Gui, J.; Feng, L.; Yang, L.; Song, L.-H. Effects of hydrogen-rich water on the nutrient composition and antioxidative characteristics of sprouted black barley. *Food Chem.* **2019**, *299*, 125095. [CrossRef] [PubMed]

94. Zhao, X.; Chen, Q.; Wang, Y.; Shen, Z.; Shen, W.; Xueqiang, Z. Hydrogen-rich water induces aluminum tolerance in maize seedlings by enhancing antioxidant capacities and nutrient homeostasis. *Ecotoxicol. Environ. Saf.* **2017**, *144*, 369–379. [CrossRef]

95. Wu, Q.; Su, N.; Cai, J.; Shen, Z.; Cui, J. Hydrogen-rich water enhances cadmium tolerance in Chinese cabbage by reducing cadmium uptake and increasing antioxidant capacities. *J. Plant Physiol.* **2015**, *175*, 174–182. [CrossRef]

96. Kolbert, Z.; Barroso, J.; Brouquisse, R.; Corpas, F.; Gupta, K.; Lindermayr, C.; Loake, G.; Palma, J.; Petřivalský, M.; Wendehenne, D.; et al. A forty year journey: The generation and roles of NO in plants. *Nitric Oxide* **2019**, *93*, 53–70. [CrossRef]

97. Arasimowicz, M.; Floryszak-Wieczorek, J. Nitric oxide as a bioactive signalling molecule in plant stress responses. *Plant Sci.* **2007**, *172*, 876–887. [CrossRef]

98. Schubert, K.R.; Evans, H.J. Hydrogen evolution: A major factor affecting the efficiency of nitrogen fixation in nodulated symbionts. *Proc. Natl. Acad. Sci. USA* **1976**, *73*, 1207–1211. [CrossRef] [PubMed]

99. Little, C.J.; Wheeler, J.A.; Sedlacek, J.; Cortés, A.J.; Rixen, C. Small-scale drivers: The importance of nutrient availability and snowmelt timing on performance of the alpine shrub *Salix herbacea*. *Oecologia* **2015**, *180*, 1015–1024. [CrossRef] [PubMed]

100. Sedlacek, J.F.; Bossdorf, O.; Cortés, A.J.; Wheeler, J.A.; Van Kleunen, M. What role do plant–soil interactions play in the habitat suitability and potential range expansion of the alpine dwarf shrub *Salix herbacea*? *Basic Appl. Ecol.* **2014**, *15*, 305–315. [CrossRef]

101. Wheeler, J.A.; Schnider, F.; Sedlacek, J.; Cortés, A.J.; Wipf, S.; Hoch, G.; Rixen, C. With a little help from my friends: Community facilitation increases performance in the dwarf shrub *Salix herbacea*. *Basic Appl. Ecol.* **2015**, *16*, 202–209. [CrossRef]

102. Li, C.; Huang, D.; Wang, C.; Wang, N.; Yao, Y.; Li, W.; Liao, W. NO is involved in H2-induced adventitious rooting in cucumber by regulating the expression and interaction of plasma membrane H+-ATPase and 14-3-3. *Planta* **2020**, *252*, 1–16. [CrossRef] [PubMed]

103. Zhu, Y.; Liao, W.; Niu, L.; Wang, M.; Ma, Z. Nitric oxide is involved in hydrogen gas-induced cell cycle activation during adventitious root formation in cucumber. *BMC Plant Biol.* **2016**, *16*, 146. [CrossRef] [PubMed]

104. Xie, Y.; Mao, Y.; Zhang, W.; Lai, D.; Wang, Q.; Shen, W. Reactive Oxygen Species-Dependent Nitric Oxide Production Contributes to Hydrogen-Promoted Stomatal Closure in Arabidopsis. *Plant Physiol.* **2014**, *165*, 759–773. [CrossRef] [PubMed]

105. Hogg, N.; Darley-Usmar, V.M.; Wilson, M.T.; Moncada, S. Production of hydroxyl radicals from the simultaneous generation of superoxide and nitric oxide. *Biochem. J.* **1992**, *281*, 419–424. [CrossRef]

106. Vandelle, E.; Delledonne, M. Peroxynitrite formation and function in plants. *Plant Sci.* **2011**, *181*, 534–539. [CrossRef] [PubMed]

107. Staszek, P.; Gniazdowska, A. Peroxynitrite induced signaling pathways in plant response to non-proteinogenic amino acids. *Planta* **2020**, *252*, 1–11. [CrossRef]

108. Kolbert, Z.; Feigl, G.; Bordé, Á.; Molnár, Á.; Erdei, L. Protein tyrosine nitration in plants: Present knowledge, computational prediction and future perspectives. *Plant Physiol. Biochem.* **2017**, *113*, 56–63. [CrossRef]

109. Alvarez, B.; Radi, R. Peroxynitrite reactivity with amino acids and proteins. *Amino Acids* **2003**, *25*, 295–311. [CrossRef] [PubMed]

110. Hanaoka, T.; Kamimura, N.; Yokota, T.; Takai, S.; Ohta, S. Molecular hydrogen protects chondrocytes from oxidative stress and indirectly alters gene expressions through reducing peroxynitrite derived from nitric oxide. *Med. Gas Res.* **2011**, *1*, 18. [CrossRef]

111. Shen, N.Y.; Bi, J.B.; Zhang, J.Y.; Zhang, S.M.; Gu, J.X.; Qu, K.; Liu, C. Hydrogen-rich water protects against inflammatory bowel disease in mice by inhibiting endoplasmic reticulum stress and promoting heme oxygenase-1 expression. *World J. Gastroenterol.* **2017**, *23*, 1375. [CrossRef]

112. Cortés, A.J.; Chavarro, C.M.; Madriñán, S.; This, D.; Blair, M.W. Molecular ecology and selection in the drought-related Asr gene polymorphisms in wild and cultivated common bean (*Phaseolus vulgaris* L.). *BMC Genet.* **2012**, *13*, 58. [CrossRef]

113. Cortés, A.J.; This, D.; Chavarro, C.; Madriñán, S.; Blair, M.W. Nucleotide diversity patterns at the drought-related DREB2 encoding genes in wild and cultivated common bean (*Phaseolus vulgaris* L.). *Theor. Appl. Genet.* **2012**, *125*, 1069–1085. [CrossRef]

114. Blair, M.W.; Cortés, A.J.; This, D. Identification of an ERECTA gene and its drought adaptation associations with wild and cultivated common bean. *Plant Sci.* **2016**, *242*, 250–259. [CrossRef] [PubMed]

115. Cortés, A.J.; Monserrate, F.A.; Ramírez-Villegas, J.; Madriñán, S.; Blair, M.W. Drought Tolerance in Wild Plant Populations: The Case of Common Beans (*Phaseolus vulgaris* L.). *PLoS ONE* **2013**, *8*, e62898. [CrossRef] [PubMed]

116. Cortés, A.J.; Blair, M.W. Genotyping by Sequencing and Genome–Environment Associations in Wild Common Bean Predict Widespread Divergent Adaptation to Drought. *Front. Plant Sci.* **2018**, *9*, 128. [CrossRef]

117. Wheeler, J.A.; Cortés, A.J.; Sedlacek, J.; Karrenberg, S.; Van Kleunen, M.; Wipf, S.; Hoch, G.; Bossdorf, O.; Rixen, C. The snow and the willows: Earlier spring snowmelt reduces performance in the low-lying alpine shrub *Salix herbacea*. *J. Ecol.* **2016**, *104*, 1041–1050. [CrossRef]

118. López-Hernández, F.; Cortés, A.J. Last-Generation Genome–Environment Associations Reveal the Genetic Basis of Heat Tolerance in Common Bean (*Phaseolus vulgaris* L.). *Front. Genet.* **2019**, *10*, 954. [CrossRef]

119. Wheeler, J.A.; Hoch, G.; Cortés, A.J.; Sedlacek, J.; Wipf, S.; Rixen, C. Increased spring freezing vulnerability for alpine shrubs under early snowmelt. *Oecologia* **2014**, *175*, 219–229. [CrossRef]

120. He, H.; He, L. Heme oxygenase 1 and abiotic stresses in plants. *Acta Physiol. Plant.* **2013**, *36*, 581–588. [CrossRef]

121. Wilks, A. Heme Oxygenase: Evolution, Structure, and Mechanism. *Antioxid. Redox Signal.* **2002**, *4*, 603–614. [CrossRef]

122. Wegiel, B.; Nemeth, Z.; Correa-Costa, M.; Bulmer, A.C.; Otterbein, L.E. Heme Oxygenase-1: A Metabolic Nike. *Antioxid. Redox Signal.* **2014**, *20*, 1709–1722. [CrossRef]

123. Rychlewski, J. Magnetic effects for the hydrogen molecule in excited states: $b^3\Sigma^+_u$ of H_2. *Mol. Phys.* **1986**, *59*, 327–336. [CrossRef]

124. Rychlewski, J. Magnetic effects for the hydrogen molecule in excited states: $B^1 \Sigma_u^+$ of H_2. *Phys. Rev. A Gen. Phys.* **1985**, *31*, 2091–2095. [CrossRef]
125. Steiner, U.E.; Ulrich, T. Magnetic field effects in chemical kinetics and related phenomena. *Chem. Rev.* **1989**, *89*, 51–147. [CrossRef]
126. Buntkowsky, G.; Walaszek, B.; Adamczyk, A.; Xu, Y.; Limbach, H.-H.; Chaudret, B. Mechanism of nuclear spin initiated *para*-H_2 to *ortho*-H_2 conversion. *Phys. Chem. Chem. Phys.* **2006**, *8*, 1929–1935. [CrossRef] [PubMed]
127. Hancock, J.T.; Hancock, T.H. Hydrogen gas, ROS metabolism, and cell signaling: Are hydrogen spin states important? *React. Oxyg. Species* **2018**, *6*, 389–395. [CrossRef]
128. Marais, A.; Adams, B.; Ringsmuth, A.K.; Ferretti, M.; Gruber, J.M.; Hendrikx, R.; Schuld, M.; Smith, S.L.; Sinayskiy, I.; Krüger, T.P.J.; et al. The future of quantum biology. *J. R. Soc. Interface* **2018**, *15*, 20180640. [CrossRef]
129. Kim, Y.; Bertagna, F.; D'Souza, E.M.; Heyes, D.J.; Johannissen, L.O.; Nery, E.T.; Pantelias, A.; Jimenez, A.S.-P.; Slocombe, L.; Spencer, M.G.; et al. Quantum Biology: An Update and Perspective. *Quantum Rep.* **2021**, *3*, 80–126. [CrossRef]
130. Huang, C.-S.; Kawamura, T.; Toyoda, Y.; Nakao, A. Recent advances in hydrogen research as a therapeutic medical gas. *Free. Radic. Res.* **2010**, *44*, 971–982. [CrossRef]
131. Wang, Y.-Q.; Liu, Y.-H.; Wang, S.; Du, H.-M.; Shen, W.-B. Hydrogen agronomy: Research progress and prospects. *J. Zhejiang Univ. Sci. B* **2020**, *21*, 841–855. [CrossRef]
132. Chuai, Y.; Gao, F.; Li, B.; Zhao, L.; Qian, L.; Cao, F.; Wang, L.; Sun, X.; Cui, J.; Cai, J. Hydrogen-rich saline attenuates radiation-induced male germ cell loss in mice through reducing hydroxyl radicals. *Biochem. J.* **2012**, *442*, 49–56. [CrossRef]
133. Halliwell, B.; Gutteridge, J.M.; Aruoma, O.I. The deoxyribose method: A simple "test-tube" assay for determination of rate constants for reactions of hydroxyl radicals. *Anal. Biochem.* **1987**, *165*, 215–219. [CrossRef]
134. Xu, G.; Chance, M.R. Hydroxyl Radical-Mediated Modification of Proteins as Probes for Structural Proteomics. *Chem. Rev.* **2007**, *107*, 3514–3543. [CrossRef]
135. El-Bahr, S.M. Biochemistry of Free Radicals and Oxidative Stress. *Sci. Int.* **2013**, *1*, 111–117. [CrossRef]
136. Tejero, I.; González-Lafont, À.; Lluch, J.M.; Eriksson, L.A. Theoretical Modeling of Hydroxyl-Radical-Induced Lipid Peroxidation Reactions. *J. Phys. Chem. B* **2007**, *111*, 5684–5693. [CrossRef] [PubMed]
137. Gilbert, B.C.; King, D.M.; Thomas, C. The oxidation of some polysaccharides by the hydroxyl radical: An e.s.r. investigation. *Carbohydr. Res.* **1984**, *125*, 217–235. [CrossRef]
138. Dizdaroglu, M.; Jaruga, P. Mechanisms of free radical-induced damage to DNA. *Free Radic. Res.* **2012**, *46*, 382–419. [CrossRef] [PubMed]
139. Gansäuer, A.; Otte, M.; Piestert, F.; Fan, C.-A. Sustainable radical reduction through catalyzed hydrogen atom transfer reactions (CHAT-reactions). *Tetrahedron* **2009**, *65*, 4984–4991. [CrossRef]

MDPI
St. Alban-Anlage 66
4052 Basel
Switzerland
Tel. +41 61 683 77 34
Fax +41 61 302 89 18
www.mdpi.com

Plants Editorial Office
E-mail: plants@mdpi.com
www.mdpi.com/journal/plants

www.ingramcontent.com/pod-product-compliance
Lightning Source LLC
LaVergne TN
LVHW070639170726
843515LV00004B/277

9 783036 550978